Lecture Notes in Mathematics

A collection of informal reports and seminars
Edited by A. Dold, Heidelberg an[illegible] Zürich

T0224658

237

Bo Stenström
University of Stockholm, Stockholm/Sweden

Rings and Modules of Quotients

Springer-Verlag
Berlin · Heidelberg · New York 1971

AMS Subject Classifications (1970): 16 A 08

ISBN 3-540-05690-4 Springer-Verlag Berlin · Heidelberg · New York
ISBN 0-387-05690-4 Springer-Verlag New York · Heidelberg · Berlin

This work is subject to copyright. All rights are reserved, whether the whole or part of the material is concerned, specifically those of translation, reprinting, re-use of illustrations, broadcasting, reproduction by photocopying machine or similar means, and storage in data banks.

Under § 54 of the German Copyright Law where copies are made for other than private use, a fee is payable to the publisher, the amount of the fee to be determined by agreement with the publisher.

© by Springer-Verlag Berlin · Heidelberg 1971. Library of Congress Catalog Card Number 70-180692. Printed in Germany.

Offsetdruck: Julius Beltz, Hemsbach

Contents

Introduction

These notes are intended to give a survey of the basic, more or less well-known, results in the theory of rings of quotients. An effort has been made to make the account as self-contained and elementary as possible. Thus we assume from the reader only a knowledge of the elements of the theory of rings and of abelian categories.

We will briefly describe the contents of the notes. Chapter 1 treats the necessary preliminaries on torsion theory. The main result here is the establishing of a 1-1 correspondence between hereditary torsion theories and topologies on a ring (Gabriel [31] and Maranda [51]).

In Chapter 2 we construct rings and modules of quotients with respect to an additive topology, following the approach of Gabriel ([10],[31]). This construction is a special instance of the construction of associated sheaves of presheaves for an additive Grothendieck topology; in these notes, however, we will not pursue that course. Rings and modules of quotients are then described in terms of injective envelopes (Johnson and Wong [114] and Lambek [46]). The main result of this chapter is the theorem of Popescu and Gabriel [62] which asserts that every Grothendieck category $\underline{C}$ is the category of modules of quotients for a suitable topology on the endomorphism ring of a generator for $\underline{C}$. The proof we give for this theorem is a simplified version of Popescu's proof [12], due to J. Lambek [48] and J.E. Roos [unpublished].

In Chapter 3 we treat some aspects of rings of quotients related to finiteness conditions. In particular we prove a theorem of Popescu and Spircu [104] which characterizes flat epimorphisms in the category of rings as a special class of rings of quotients. In this context we also discuss rings of fractions, i.e. rings obtained by inverting elements of some multiplicatively closed subset of a ring.

Chapter 4 contains some material on self-injective rings. Various well-known characterizations of quasi-Frobenius rings are given here. These results are used in Chapter 5 where we discuss maximal rings of quotients and classical rings of quotients. Necessary and sufficient conditions for these rings to be regular, semi-simple or quasi-Frobenius are given (results due to Gabriel [31], Goldie [91], Mewborn and Winton [54], Sandomierski [68], and others).

The references at the end of each section are intended to tell the reader where he can find a further discussion of the treated topics. Their purpose is not to record credit for the results of the section.

I am grateful to J.E. Roos for allowing me to use his preliminary manuscripts for [66].

Stockholm, December 1970

Bo Stenström

Some notation

All rings have identity elements. A denotes a ring and all modules are right A-modules, unless otherwise is stated. The category of right A-modules is denoted by Mod-A, and we write M_A to indicate that M is in this category.

$E(M)$ is the injective envelope of M_A.

If L is a submodule of M and $x \in M$, then $(L:x) = \{a \in A \mid xa \in L\}$.

$J(A)$, or simply J, denotes the Jacobson radical of A.

Chapter 1. Torsion theory

§ 1. Preradicals.

A preradical r of Mod-A assigns to each module M a submodule $r(M)$ in such a way that every homomorphism $M \to N$ induces $r(M) \to r(N)$ by restriction. In other words, a preradical is a subfunctor of the identity functor of Mod-A .

The class of all preradicals of Mod-A is a complete lattice, because there is a partial ordering in which $r_1 \leq r_2$ means that $r_1(M) \subset r_2(M)$ for all modules M , and any family (r_i) of preradicals has a supremum Σr_i and an infimum $\bigcap r_i$, defined in the obvious ways. If r_1 and r_2 are preradicals, one defines preradicals $r_1 r_2$ and $r_1 : r_2$ as $r_1 r_2(M) = r_1(r_2(M))$ and $(r_1 : r_2)(M)/r_1(M) = r_2(M/r_1(M))$. A preradical r is called idempotent if $rr = r$ and is called a radical if $r:r = r$, i.e. if $r(M/r(M)) = 0$ for every module M .

Proposition 1.1. For every preradical r there exists a largest idempotent preradical $\hat{r}$ smaller than r , and there exists a smallest radical $\bar{r}$ larger than r .

Proof. To obtain $\hat{r}$, we define a sequence of preradicals by transfinite induction as follows: if β is not a limit ordinal we put $r^{\beta} = rr^{\beta-1}$, and if β is a limit ordinal we put $r^{\beta} = \bigcap_{\alpha < \beta} r^{\alpha}$. In this way we obtain a decreasing sequence of preradicals r^{β} and we put $\hat{r} = \bigcap_{\beta} r^{\beta}$. It is easy to see that the preradical $\hat{r}$ is idempotent, and that it in fact is the largest idempotent preradical smaller than r .

$\bar{r}$ is obtained in a dual fashion. If β is not a limit ordinal we set $r_{\beta} = r_{\beta-1} : r$, i.e. r_{β} is given by $r_{\beta}(M)/r_{\beta-1}(M) = r(M/r_{\beta-1}(M))$, and for a limit ordinal β we set $r_\beta = \sum_{\alpha<\beta} r_\alpha(M)$. This gives an ascending sequence of preradicals r_β , and we set $\bar{r} = \sum_\beta r_\beta$.

Lemma 1.2. If r is a radical and $L \subset r(M)$, then $r(M/L) = r(M)/L$.

Proof. The canonical homomorphism $M \to M/L$ induces $r(M) \to r(M/L)$ with kernel $r(M) \cap L = L$, so $r(M)/L \subset r(M/L)$. On the other hand, the canonical homomorphism $M/L \to M/r(M)$ maps $r(M/L)$ into zero, so $r(M/L) \subset r(M)/L$.

Proposition 1.3. (i) If r is idempotent, then so is also $\bar{r}$.

(ii) If r is a radical, then so is also $\hat{r}$.

Proof. To prove (i) we have to show:

(1) If r_1 and r_2 are idempotent, then so is also $r_1 : r_2$.

(2) If r_i $(i \in I)$ are idempotent, then so is also $\sum r_i$.

For (ii) we have to show:

(3) If r_1 and r_2 are radicals, then so is also $r_1 r_2$.

(4) If r_i $(i \in I)$ are radicals, then so is also $\bigcap r_i$.

(1): We have $r_1(M) \subset (r_1 : r_2)(M) \subset M$, and since r_1 is idempotent this gives $r_1(M) \subset r_1(r_1 : r_2)(M) \subset r_1(M)$, so $r_1(r_1 : r_2) = r_1$. Therefore we obtain from $(r_1 : r_2)(r_1 : r_2)(M)/r_1(r_1 : r_2)(M) = r_2((r_1 : r_2)(M)/r_1(r_1 : r_2)(M))$ that $(r_1 : r_2)(r_1 : r_2)(M)/ r_1(M) =$ $= r_2((r_1 : r_2)(M)/r_1(M)) = r_2 r_2(M/r_1(M)) = r_2(M/r_1(M)) = (r_1 : r_2)(M)/r_1(M)$. Consequently $r_1 : r_2$ is idempotent.

(2): From $r_i(M) \subset \sum_j r_j(M) \subset M$ we get $r_i(M) = r_i(\sum_j r_j(M))$, so $(\sum_i r_i)(\sum_j r_j)(M) = \sum_i r_i(\sum_j r_j(M)) = \sum r_i(M)$.

(3): Since $r_1 r_2(M) \subset r_2(M)$, we get from Lemma 1.2 that $r_1 r_2(M/r_1 r_2(M)) = r_1(r_2(M)/r_1 r_2(M)) = 0$.

(4): Since $\bigcap_j r_j(M) \subset r_i(M)$, we get from Lemma 1.2 that
$\bigcap_i r_i(M/\bigcap_j r_j(M)) = \bigcap_i r_i(M)/\bigcap_j r_j(M) = 0$.

Proposition 1.4. The following assertions are equivalent for a preradical r :

(a) r is a left exact functor.

(b) If $L \subset M$, then $r(L) = r(M) \cap L$.

The proof is easy. A consequence of this result is that a left exact preradical always is idempotent.

To each preradical r we associate two classes of modules, namely

$\underline{\underline{T}}_r = \{M \mid r(M) = M\}$

$\underline{\underline{F}}_r = \{M \mid r(M) = 0\}$.

Clearly:

Proposition 1.5. $\underline{\underline{T}}_r$ is closed under quotient modules, while $\underline{\underline{F}}_r$ is closed under submodules.

Corollary 1.6. If $M \in \underline{\underline{T}}_r$ and $N \in \underline{\underline{F}}_r$, then $Hom_A(M,N) = 0$.

Examples:

1. Let A be an integral domain. For every module M we let $t(M)$ denote the torsion submodule of M , and let $d(M)$ denote the maximal divisible submodule of M . Both t and d are idempotent radicals. t is also left exact, which is not the case for d .

2. Let A be an arbitrary ring. For each right module M we let $s(M)$ denote the socle of M , i.e. the sum of all simple submodules of M , and let $r(M)$ denote the intersection of all maximal proper submodules of M . s is a left exact preradical, while r is a radical.

Exercises:

1. Show that every preradical commutes with direct sums, and that a left exact preradical commutes with directed unions of submodules.
2. Suppose r is an idempotent preradical. Show that $\bar{r}(M)$ is the largest submodule L of M such that $r(L/L') \neq 0$ for every $L' \subsetneq L$.
3. Show that if r is a left exact preradical, then also $\bar{r}$ is left exact.
4. Show that a preradical r is left exact if and only if r is idempotent and $\underline{\underline{T}}_r$ is closed under submodules.

References: Radicals have been studied by many authors, usually in the context of general categories or the category of rings. The case of module categories has been treated e.g. by Amitsur [4], Kurosch [44] and Maranda [51]. For some further references to the russian literature, see Math. Reviews 35, no. 234.

§ 2. Torsion theories

Definition. A torsion theory for Mod-A is a pair $(\underline{\underline{T}},\underline{\underline{F}})$ of classes of right A-modules such that

(i) $\operatorname{Hom}(T,F) = 0$ for all $T \in \underline{\underline{T}}$, $F \in \underline{\underline{F}}$;

(ii) $\underline{\underline{T}}$ and $\underline{\underline{F}}$ are maximal classes having property (i).

The modules in $\underline{\underline{T}}$ are called torsion modules and the modules in $\underline{\underline{F}}$ are torsion-free.

Any given class $\underline{C}$ of modules <u>generates</u> a torsion theory in in the following way:

$\underline{F} = \{F \mid \mathrm{Hom}(C,F) = 0 \text{ for all } C \in \underline{C}\}$,

$\underline{T} = \{T \mid \mathrm{Hom}(T,F) = 0 \text{ for all } F \in \underline{F}\}$.

Clearly this $(\underline{T}, \underline{F})$ is a torsion theory, and we also note that $\underline{T}$ is the smallest class of torsion modules containing $\underline{C}$.

Dually, the class $\underline{C}$ <u>cogenerates</u> a torsion theory $(\underline{T}, \underline{F})$ such that $\underline{F}$ is the smallest class of torsion-free modules containing $\underline{C}$.

We can characterize those classes of modules which may appear as the class of torsion modules for some torsion theory; such a class of modules will be called a <u>torsion class</u> . (Classes will henceforth be assumed to be closed under isomorphisms).

<u>Proposition 2.1</u>. A class of modules is a torsion class if and only if it is closed under quotients, direct sums and extensions.

<u>Proof</u>. A class $\underline{C}$ is said to be "closed under extensions" if for every exact sequence $0 \to L \to M \to N \to 0$ with L and N in $\underline{C}$, also $M \in \underline{C}$.

Suppose $(\underline{T}, \underline{F})$ is a torsion theory. $\underline{T}$ is obviously closed under quotients and direct sums, since $\mathrm{Hom}(\oplus T_i, F) = \prod \mathrm{Hom}(T_i, F)$. Let $0 \to L \to M \to N \to 0$ be exact with L and N in $\underline{T}$. If F is torsion-free and $\alpha: M \to F$, then α is zero on L , so α factors over N . But also $\mathrm{Hom}(N,F) = 0$, so $\alpha = 0$. Hence $M \in \underline{T}$.

Conversely, assume $\underline{C}$ to be closed under quotients, direct sums and extensions. Let $(\underline{T}, \underline{F})$ be the torsion theory generated by $\underline{C}$. We want to show that $\underline{C} = \underline{T}$, so suppose $\mathrm{Hom}(T, F) = 0$ for all $F \in \underline{F}$. Let C be the sum of all submodules of T

belonging to $\underline{C}$. C then also belongs to $\underline{C}$, since $\underline{C}$ is closed under direct sums and quotients. To show that $C = T$, it suffices to show that $T/C \in \underline{\underline{F}}$. Suppose we have $\alpha : C' \to T/C$ where $C' \in \underline{C}$. The image of α is a module in $\underline{C}$, and if $\alpha = 0$, then we would get a submodule of T which strictly contains C and belongs to $\underline{C}$, since $\underline{C}$ is closed under extensions. This would contradict the maximality of C , and thus we must have $\alpha = 0$, and $T/C \in \underline{\underline{F}}$.

Dually one shows (by working in an abelian category one obtains this immediately by duality):

<u>Proposition 2.2.</u> A class of modules is a torsion-free class if and only if it is closed under submodules, direct products and extensions.

Let $(\underline{\underline{T}}, \underline{\underline{F}})$ be a torsion theory. It follows from 2.1 that every module M has a unique largest submodule $t(M)$ belonging to $\underline{\underline{T}}$, called the <u>torsion submodule</u> of M . In this way we obtain a preradical t of Mod-A , and it is easily verified that t is an idempotent radical. Conversely, given any idempotent radical t of Mod-A , the pair of classes $(\underline{\underline{T}}_t, \underline{\underline{F}}_t)$ defined in § 1 is a torsion theory. As a result we obtain:

<u>Proposition 2.3.</u> There is a 1-1 correspondence between torsion theories and idempotent radicals.

<u>Proposition 2.4.</u> Let r be an idempotent preradical. $\bar{r}$ is the idempotent radical associated with the torsion theory generated by $\underline{\underline{T}}_r$.

<u>Proof.</u> We know from 1.1 and 1.3 that $\dot{\bar{r}}$ is the smallest idempotent

radical containing r . It must therefore correspond to the smallest torsion class containing $\underline{\underline{T}}_r$, which is just $\underline{\underline{T}}_{\bar{r}}$.

Proposition 2.5. Let $\underline{C}$ be a class of modules closed under quotients. The torsion class generated by $\underline{C}$ consists of all modules M such that each non-zero quotient of M has a non-zero submodule in $\underline{C}$.

Proof. Let $T(\underline{C})$ be the class of modules M satisfying the announced condition. $T(\underline{C})$ obviously contains $\underline{C}$, and if $\underline{\underline{T}}$ is a torsion class containing $\underline{C}$, then it is easily seen that $\underline{\underline{T}} \supset T(\underline{C})$. It therefore remains for us to show that $T(\underline{C})$ is a torsion class. $T(\underline{C})$ is clearly closed under quotients. Suppose (M_i) is a family of modules in $T(\underline{C})$. If $\alpha: \oplus M_i \to N$ is a non-zero epimorphism, choose M_i such that α is non-zero on M_i . Since $\alpha(M_i)$ contains a non-zero submodule in $\underline{C}$, so does also N . $T(\underline{C})$ is thus closed under direct sums.

Let $0 \to L \to M \to N \to 0$ be exact with L , N in $T(\underline{C})$. Suppose $\alpha: M \to M'$ is a non-zero epimorphism. We get an exact commutative diagram

$$\begin{array}{ccccccccc} & & 0 & & 0 & & 0 & & \\ & & \downarrow & & \downarrow & & \downarrow & & \\ 0 & \to & K \cap L & \to & K & \to & K' & \to & 0 \\ & & \downarrow & & \downarrow & & \downarrow & & \\ 0 & \to & L & \to & M & \to & N & \to & 0 \\ & & \downarrow & & \downarrow & & \downarrow & & \\ 0 & \to & L/K \cap L & \to & M' & \to & N' & \to & 0 \\ & & \downarrow & & \downarrow & & \downarrow & & \\ & & 0 & & 0 & & 0 & & \end{array}$$

If $K \cap L \neq L$, then $L/K \cap L$ has a non-zero submodule in $\underline{C}$ and so has then also M' . If $L \subset K$, then $M' \cong N'$ and so M' still has a non-zero submodule in $\underline{C}$. Hence $M \in T(\underline{C})$.

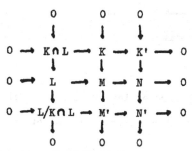

Of particular interest for us will be the torsion theories $(\mathbf{T}, \mathbf{F})$ where $\mathbf{T}$ is closed under submodules.

<u>Proposition 2.6</u>. Let $(\mathbf{T}, \mathbf{F})$ be a torsion theory and let t be the corresponding radical. The following assertions are equivalent:

(a) $\mathbf{T}$ is closed under submodules.

(b) $\mathbf{F}$ is closed under injective envelopes.

(c) t is left exact.

(d) If $L \subset M$, then $t(L) = L \cap t(M)$.

<u>Proof</u>. (c) $\Leftrightarrow$ (d) was remarked in 1.4.

(a) $\Rightarrow$ (d): If $L \subset M$, then certainly $t(L) \subset L \cap t(M)$. But $L \cap t(M)$ is a torsion module, since it is a submodule of $t(M)$, so we must have $t(L) = L \cap t(M)$.

(d) $\Rightarrow$ (b): Suppose $F \in \mathbf{F}$. Then $t(E(F)) \cap F = t(F) = 0$. But F is essential in $E(F)$, so this implies $t(E(F)) = 0$.

(b) $\Rightarrow$ (a): Suppose $T \in \mathbf{T}$ and $C \subset T$. There is a commutative diagram

$$\begin{array}{ccc} C & \hookrightarrow & T \\ {\scriptstyle\alpha}\downarrow & & \downarrow{\scriptstyle\beta} \\ C/t(C) & \hookrightarrow & E(C/t(C)) \end{array}$$

But $E(C/t(C))$ is torsion-free, so $\beta = 0$. This implies $\alpha = 0$ and hence $C = t(C) \in \mathbf{T}$.

A torsion theory with the properties stated in 2.6 is called <u>hereditary</u>. Applying 2.6 to Proposition 2.3, we obtain:

<u>Corollary 2.7</u>. There is a 1-1 correspondence between hereditary torsion theories and left exact radicals.

<u>Proposition 2.8.</u> A hereditary torsion theory is generated by the family of those cyclic modules A/I which are torsion modules.

<u>Proof.</u> A module M is torsion-free if and only if $\mathrm{Hom}(A/I,M) = 0$ for all $A/I \in \underline{\underline{T}}$.

A hereditary torsion theory is thus uniquely determined by the family of right ideals I for which A/I is a torsion module. In the next § we will characterize these families of right ideals.

<u>Proposition 2.9.</u> Let $\underline{C}$ be a class of modules closed under quotients and submodules. The torsion theory generated by $\underline{C}$ is hereditary.

<u>Proof.</u> We will show that the class of torsion-free modules is closed under injective envelopes. Suppose F is torsion-free and there exists a non-zero $\alpha : C \to E(F)$ with $C \in \underline{C}$. Then $\mathrm{Im}\,\alpha \in \underline{C}$ and $F \cap \mathrm{Im}\,\alpha$ is a non-zero submodule of F belonging to $\underline{C}$, which is a contradiction.

<u>Examples:</u>

1. Let A be an integral domain. The "standard" torsion theory is hereditary. An example of a non-hereditary torsion theory is given by $\underline{\underline{T}} = \{\text{divisible modules}\}$, $\underline{\underline{F}} = \{\text{reduced modules}\}$. The next two examples generalize the standard torsion theory to arbitrary rings.

2. Let A be an arbitrary ring and let $\underline{C}$ be the class of modules of the form M/L where L is an essential submodule of M. The torsion theory generated by $\underline{C}$ is called the <u>Goldie torsion theory</u>. Since $\underline{C}$ is closed under quotients and submodules, the Goldie torsion theory is hereditary and is generated by the

family of cyclic modules A/I where I is an essential right ideal of A. M is a Goldie torsion module if and only if each non-zero quotient of M has a submodule $\neq 0$ in $\underline{C}$ which we may assume cyclic (2.5). Hence:

Proposition 2.10. M is a Goldie torsion module if and only if for each submodule $L \subsetneq M$ there exists $0 \neq x \in M$ such that $(L:x)$ is an essential right ideal.

3. The torsion theory cogenerated by $E(A)$ will be called the Lambek torsion theory. It is hereditary, because we have:

Proposition 2.11. M is a Lambek torsion module if and only if $\mathrm{Hom}(L,A) = 0$ for every submodule L of M.

Proof. If $L \subset M$ and $\alpha: L \to A$ is a non-zero homomorphism, then α may be extended to $M \to E(A)$. So every Lambek torsion module has the mentioned property. The converse follows from the fact that if $\alpha: M \to E(A)$ is a non-zero homomorphism, then $\mathrm{Im}\,\alpha \cap A \neq 0$, so $\alpha^{-1}(A) \to A$ is non-zero.

Exercises:

1. Show that in the definition of a torsion theory $(\underline{\underline{T}}, \underline{\underline{F}})$ one may replace the axiom (ii) by:

   (ii)': Each module M has a submodule $T \in \underline{\underline{T}}$ such that $M/T \in \underline{\underline{F}}$; moreover $\underline{\underline{T}}$ and $\underline{\underline{F}}$ are closed under isomorphisms.

2. Let $\underline{C}$ be any class of modules. Let $T(\underline{C})$ be the class of of modules M such that each non-zero quotient module of M has a non-zero submodule in $\underline{C}$. Show that:

(i) $T(\underline{C})$ is a torsion class;

(ii) if $\underline{\underline{T}}$ is a torsion class containing $\underline{C}$, then $\underline{\underline{T}} \supset T(\underline{C})$ (note that $\underline{C} \not\subset T(\underline{C})$ in general).

3. Let $\underline{\underline{T}}$ be a hereditary torsion class, generated by a class $\underline{C}$ closed under quotients and submodules. Show that if $\underline{\underline{T}}'$ is a hereditary torsion class contained in $\underline{\underline{T}}$, then $\underline{\underline{T}}'$ is generated by $\underline{C} \cap \underline{\underline{T}}'$. (Hint: use 2.5).

4. (i) State and prove the dual of Proposition 2.9.

(ii) Show that a hereditary torsion theory $(\underline{\underline{T}}, \underline{\underline{F}})$ is cogenerated by the family of injective modules $E(A/I)$ such that $A/I \in \underline{\underline{F}}$.

(iii) Let E be an injective module. Show that the torsion theory cogenerated by $\{E\}$ is hereditary.

5. A module M is called <u>torsionless</u> if for every $0 \neq x \in M$ there exists $\alpha: M \to A$ such that $\alpha(x) \neq 0$.

(i) Show that M is torsionless if and only if M is a submodule of a direct product of copies of A.

(ii) Show that the class $\underline{\underline{L}}$ of torsionless modules is closed under submodules and direct products.

(iii) Let $\underline{\underline{T}}$ be the class of modules M for which $\mathrm{Hom}(M,A) = 0$. Show that the following assertions are equivalent:

(a) $E(A)$ is torsionless.

(b) $\underline{\underline{L}}$ is closed under injective envelopes.

(c) $(\underline{\underline{T}}, \underline{\underline{L}})$ is a hereditary torsion theory.

(d) $(\underline{\underline{T}}, \underline{\underline{L}})$ is the Lambek torsion theory.

If direct products of projective modules are projective, show that these conditions are equivalent to:

(e) $E(A)$ is projective.

(Hint: show (a)$\Rightarrow$(b)$\Rightarrow$(c)$\Rightarrow$(d)$\Rightarrow$(a). To show (b)$\Rightarrow$(c), suppose given $0 \to L \to M \to M/L \to 0$ with L , M/L in $\underline{L}$; choose maximal K with $K \cap L = 0$ and remark that $K \subset M/L$ and $K + L$ is essential in M).

References: Alin [1], Dickson [21], Mishina and Skornjakov [55]. Exercise 5 is based on a paper by Wu, Mochizuki and Jans [82].

§ 3. Topologies

In this § we will establish a 1-1 correspondence between hereditary torsion theories and certain families of right ideals, called "topologies".

Definition. A right additive topology (or just topology) is a non-empty family $\underline{F}$ of right ideals of A , satisfying:

T 1. If $I \in \underline{F}$ and $a \in A$, then $(I:a) \in \underline{F}$.

T 2. If I is a right ideal and there exists $J \in \underline{F}$ such that $(I:a) \in \underline{F}$ for every $a \in J$, then $I \in \underline{F}$.

Note that T 1 , together with the assumption that $\underline{F}$ is non-empty, implies $A \in \underline{F}$.

Lemma 3.1. A topology $\underline{F}$ is a filter, i.e. satisfies:

T 3. If $J \in \underline{F}$ and $J \subset I$, then $I \in \underline{F}$.

T 4. If I and J are in $\underline{F}$, then $I \cap J \in \underline{F}$.

Proof. T 3: If $a \in J$, then $(I:a) = A \in \underline{F}$, so $I \in \underline{F}$ by T 2.
T 4: If $a \in J$, then $((I \cap J):a) = (I:a) \cap (J:a) = (I:a) \in \underline{F}$ by T 1, so $I \cap J \in \underline{F}$ by T 2.

The axioms T 1, 3 and 4 state that A is a topological ring in which $\underline{F}$ is a fundamental system of neighborhoods of zero. A non-empty family of right ideals satisfying T 1, 3 and 4 is called a pretopology.

Proposition 3.2. There is a 1-1 correspondence between:

1) Non-empty families of right ideals of A satisfying T 1.
2) Non-empty classes $\underline{C}$ of modules such that $M \in \underline{C}$ if and only if every cyclic submodule of M is in $\underline{C}$.

Proof. If $\underline{F}$ is a family of right ideals satisfying T 1, we define $\underline{C}$ as the class of modules M for which $\mathrm{Ann}(x) \in \underline{F}$ for every $x \in M$. $\underline{C}$ obviously satisfies (2). Conversely, if $\underline{C}$ is the given class of modules, we put $\underline{F} = \{ I \mid A/I \in \underline{C} \}$. This $\underline{F}$ satisfies T 1, for if $I \in \underline{F}$ and $a \in A$, then left multiplication by a induces an exact sequence

$$0 \longrightarrow (I:a) \longrightarrow A \longrightarrow A/I$$

which shows that $A/(I:a) \subset A/I$.

It remains to verify that these correspondences are the inverses of each other. Starting with $\underline{F}$ we get $\underline{C} = \{ M \mid \mathrm{Ann}(x) \in \underline{F}$ for all $x \in M \}$, and from this we get $\{ I \mid A/I \in \underline{C} \} = \{ I \mid (I:a) \in \underline{F}$ for all $a \in A \} = \underline{F}$ by T 1. On the other hand, if we start with $\underline{C}$, we first get $\underline{F} = \{ I \mid A/I \in \underline{C} \}$ and then $\{ M \mid \mathrm{Ann}(x) \in \underline{F} \} = \{ M \mid \text{each cyclic submodule} \in \underline{C} \} = \underline{C}$.

Proposition 3.3. There is a 1-1 correspondence between:

1) Pretopologies on A.
2) Classes of modules closed under submodules, quotients and direct sums.
3) Left exact preradicals of Mod-A.

Proof. (1) $\leftrightarrow$ (2): This is a particular case of 3.2. For it is easy to see that if $\underline{F}$ is a pretopology, then the class of modules M for which $Ann(x) \in \underline{F}$, all $x \in M$, is closed under submodules (by 3.2), quotients (by T 3) and direct sums (by T 4). Conversely, if $\underline{C}$ is the given class of modules, then $\underline{F} = \{ I \mid A/I \in \underline{C} \}$ is a pretopology, for T 1 is satisfied by 3.2, T 3 is obviously satisfied and so is also T 4 because $A/I \cap J$ is a submodule of $A/I \oplus A/J$.

(2) $\leftrightarrow$ (3): If $\underline{C}$ is the given class of modules, we let $r(M)$ be the largest submodule of M belonging to $\underline{C}$. r is a left exact preradical. Conversely, if r is a left exact preradical, then the class $\underline{\underline{T}}_r$ is closed under quotients and submodules. It is also closed under direct sums, for if $M_i \in \underline{\underline{T}}_r$, then $r(\oplus M_i) \cap M_j = r(M_j) = M_j$ for each j, so $r(\oplus M_i) = \oplus M_i$ (or use § 1, Exercise 1). The correspondence between 2 and 3 is clearly 1-1.

Note in particular that given the pretopology $\underline{F}$, the preradical r is obtained as $r(M) = \{ x \in M \mid Ann(x) \in \underline{F} \}$.

Theorem 3.4. There is a 1-1 correspondence between:

1) Topologies on A.

2) Hereditary torsion theories for Mod-A.

3) Left exact radicals of Mod-A.

Proof. We have already established the 1-1 correspondence (2) $\leftrightarrow$ (3) in Corollary 2.7. If $\underline{F}$ is a topology, then the corresponding left exact preradical t, defined above as $t(M) = \{ x \in M \mid Ann(x) \in \underline{F} \}$, is a radical. In fact, if $\bar{x} \in M/t(M)$ and $Ann(\bar{x}) \in \underline{F}$, then $Ann(\bar{x}) = \{ a \mid xa \in t(M) \} = \{ a \mid Ann(xa) \in \underline{F} \} = \{ a \mid (Ann(x):a) \in \underline{F} \}$, and it follows from T 2 that $Ann(x) \in \underline{F}$ and hence $\bar{x} = 0$.

On the other hand, if $(\underline{T}, \underline{F})$ is a hereditary torsion theory, then the corresponding pretopology $\underline{F} = \{ I \mid A/I \in \underline{T} \}$ satisfies T 2. For if I is a right ideal such that $(I:a) \in \underline{F}$ for all $a \in J$ for some $J \in \underline{F}$, we consider the exact sequence

$$0 \longrightarrow J/I \cap J \longrightarrow A/I \longrightarrow A/I \cup J \longrightarrow 0$$

where $A/I \cup J \in \underline{T}$ since it is a quotient module of $A/J \in \underline{T}$, and also $J/I \cap J \in \underline{T}$ since $a \in J$ implies $((I \cap J):a) = (I:a) \in \underline{F}$. Since $\underline{T}$ is closed under extensions, it follows that $A/I \in \underline{T}$ and hence $I \in \underline{F}$. This concludes the proof of the theorem.

If $\underline{F}_1$ and $\underline{F}_2$ are topologies on A, we say that $\underline{F}_1$ is <u>weaker</u> than $\underline{F}_2$ (and $\underline{F}_2$ is <u>stronger</u> than $\underline{F}_1$) if $\underline{F}_1 \subset \underline{F}_2$. It is clear that any intersection of topologies is a topology, so if $\underline{E}$ is any family of right ideals of A, there exists a weakest topology $J(\underline{E})$ containing $\underline{E}$. The correspondence between pretopologies and classes of modules described in 3.3 preserves inclusions. It follows in particular that if $\underline{E}$ is a pretopology, then $J(\underline{E})$ is the topology corresponding to the hereditary torsion theory generated by $\{ A/I \mid I \in \underline{E} \}$.

<u>Examples</u>:

1. The family $\underline{E}$ of essential right ideals is a pretopology, but does not always satisfy T 2. The corresponding left exact preradical is usually denoted by Z, and one calls $Z(M)$ the <u>singular submodule</u> of M. The hereditary torsion theory

corresponding to $J(\underline{E})$ is the Goldie torsion theory (§ 2, Example 2). The Goldie torsion radical G is the smallest radical containing Z. The transfinite process which leads from Z to G (Proposition 1.1) gets stationary at an early stage; in fact we have, in the notation of 1.1:

<u>Proposition 3.5</u>. $G = Z_2$.

<u>Proof</u>. We first remark that $Z(M)$ is an essential submodule of $G(M)$ for every module M. For if $L \subset G(M)$ and $L \cap Z(M) = 0$, then $Z(L) = L \cap Z(M) = 0$ and L is thus torsion-free, which implies $L = 0$. We now have $Z_2(M)/Z(M) = Z(M/Z(M)) \supset G(M)/Z(M)$, so $Z_2(M) = G(M)$.

<u>Proposition 3.6</u>. If $\underline{E}$ is the family of essential right ideals, then $J(\underline{E}) = \{ I \mid \text{there exists } J \in \underline{E} \text{ such that } I \subset J \text{ and } (I:a) \in \underline{E} \text{ for all } a \in J \}$.

<u>Proof</u>. If I satisfies the condition, then $I \in J(\underline{E})$ by T 2. If $I \in J(\underline{E})$, then A/I is a Goldie torsion module, so $A/I = Z_2(A/I)$. We write $Z(A/I) = J/I$. The formula defining Z_2 gives $A/J = Z(A/J)$ and hence $J \in \underline{E}$. We also have $Z(J/I) = Z^2(A/I) = Z(A/I) = J/I$, so $(I:a) \in \underline{E}$ for all $a \in J$.

2. The right ideals of the topology corresponding to the Lambek torsion theory are called <u>dense</u>. We denote this toplogy by $\underline{D}$.

<u>Proposition 3.7</u>. $\underline{D}$ is the strongest topology on A such that A is torsion-free.

<u>Proof</u>. A is trivially torsion-free for $\underline{D}$. If $(\underline{\underline{T}}, \underline{\underline{F}})$ is any hereditary torsion theory such that $A \in \underline{\underline{F}}$, then $\underline{\underline{T}}$ contains the Lambek torsion modules and the corresponding topology is weaker than $\underline{D}$.

A more explicit description of $\underline{D}$ is given by:

<u>Proposition 3.8</u>. A right ideal I is dense if and only if $(I:a)$ has no left annihilators for any $a \in A$.

<u>Proof</u>. Suppose $(I:a)$ never has left annihilators. If there exists a non-zero homomorphism $A/I \to E(A)$, then there exists $0 \neq x \in E(A)$ such that $xI = 0$. Let $a \in A$ be such that $0 \neq xa \in A$. If $b \in (I:a)$, then $xab = 0$, and hence xa is a left annihilator of $(I:a)$, a contradiction.

Suppose conversely that I is dense, and suppose there exist a, b in A with $b(I:a) = 0$. We get a commutative diagram

$$\begin{array}{ccc} A/(I:a) & \xrightarrow{\ f\ } & A \\ {\scriptstyle g}\downarrow & & \downarrow \\ A/I & \xrightarrow[\ h\ ]{} & E(A) \end{array}$$

where $f(\overline{c}) = bc$, $g(\overline{c}) = ac$ and h exists because $E(A)$ is injective. I dense implies $h = 0$ and hence $b = 0$.

<u>Corollary 3.9</u>. Every dense right ideal is essential in A.

The ring A is called <u>right non-singular</u> if the singular right ideal $Z(A)$ is zero.

<u>Proposition 3.10</u>. If A is right non-singular, then every essential right ideal is dense, and the Lambek torsion theory coincides with the Goldie torsion theory.

<u>Proof</u>. If I is an essential right ideal, then also $(I:a)$ is essential for every $a \in A$, and so non-singularity implies that $(I:a)$ can have no non-zero left annihilators.

3. Let $\underline{F}$ be the family of those right ideals which contain a non-zero-divisor. $\underline{F}$ is a topology if and only if A satisfies the "right Ore condition", i.e. for every non-zero-divisor s and arbitrary $a \in A$ there exist a non-zero-divisor t and $b \in A$ such that $at = sb$.

4. Let S be a multiplicatively closed subset of A (by this we also understand that $1 \in S$). The family $\underline{F}(S) = \{$ right ideals $I \mid$ for every $a \in A$ there exists $s \in S$ with $as \in I \}$ is easily verified to be a topology.

5. Let $\underline{C}$ be the class of semi-simple modules, which is closed under submodules, quotients and direct sums. The corresponding pretopology consists of all finite intersections of maximal right ideals, and the corresponding left exact preradical s associates to each module M its <u>socle</u> $s(M)$. The corresponding torsion radical is as usual denoted by $\bar{s}$. Note that the torsion-free modules are those modules which have zero socle.

<u>Proposition 3.11</u>. When A is a right noetherian ring, $\bar{s}(M)$ is the sum of all artinian submodules of M .

<u>Proof</u>. A module is called artinian if it satisfies DCC (descending chain condition) on submodules. Put $t(M)$ = sum of all artinian submodules of M . t is clearly a left exact preradical. It is not quite so obvious that it is a radical, and here we need the noetherian hypothesis. We write $t(M/t(M)) = L/t(M)$, and want to show that every finitely generated submodule L' of L is artinian. We have $L'/t(L') = L'/L' \cap t(M) = L' + t(M)/t(M) \subset$ $\subset L/t(M)$ and hence $L'/t(L')$ is artinian. Since A is noetherian,

also $t(L')$ is finitely generated and hence artinian. Since the class of artinian modules is closed under extensions, it follows that L' is artinian.

t thus defines a torsion theory, and this torsion theory is the same as the one given by $\bar{\bar{s}}$, because the torsion-free modules are clearly the same in both cases.

The ring A is called <u>right semi-artinian</u> if every non-zero module has non-zero socle. In that case we have no torsion-free modules $\neq 0$, and thus every module is a torsion module, i.e. $\bar{\bar{s}}(M) = M$ for all M. In particular we see from 3.11 that a right noetherian and semi-artinian ring is right artinian.

<u>Proposition 3.12</u>. A is right semi-artinian if and only if every hereditary torsion theory is generated by a class of simple modules.

<u>Proof</u>. If A is right semi-artinian, then the torsion class generated by all simple modules is Mod-A. It follows from § 2, Exercise 3 that any hereditary torsion theory is generated by a class of simple modules.

The converse follows from the trivial fact that Mod-A is a hereditary torsion class.

<u>Exercises</u>:

1. Show that if $\underline{F}$ is a topology and $I, J \in \underline{F}$, then also $IJ \in \underline{F}$.
2. Suppose the ring A is commutative and $\underline{F}$ is a pretopology such that every ideal in $\underline{F}$ contains a finitely generated ideal in $\underline{F}$. Show that if the product of two ideals in $\underline{F}$

always is in $\underline{F}$, then $\underline{F}$ is a topology.

3. (i) Show that A (as a right A-module) is a Goldie torsion module if and only if Z(A) is an essential right ideal in A .
   (ii) Show that the ring $\underline{Z}/p^2\underline{Z}$ (where p is a prime number) is a Goldie torsion module over itself.
4. (i) Let I be a right ideal. Show that the family of right ideals containing I is a topology if and only if I is a two-sided ideal and $I^2 = I$.
   (ii) Show that the following assertions are equivalent for a hereditary torsion class $\underline{\underline{T}}$:
   (a) $\underline{\underline{T}}$ is closed under direct products.
   (b) The corresponding topology is of the form $\{J \mid J \supset I\}$ for some right ideal I .
   (c) $\underline{\underline{T}} = \{M \mid MI = 0\}$ for some ideal I .

References: Alin [1] , Bourbaki [10] (p. 157), Gabriel [31] (p. 411), Maranda [51] , Mishina and Skornjakov [55], Roos [66] (ch. 1).

§ 4. Stable torsion theories

In this and the following sections of this chapter, we will consider some more particular aspects of torsion. The results of these sections are not necessary for the understanding of the following chapters.

A hereditary torsion theory $(\underline{\underline{T}}, \underline{\underline{F}})$ is called stable if $\underline{\underline{T}}$ is closed under injective envelopes. Alternative characterizations of stable torsion may be given by using the concept of "complemented" (or "closed") submodules. A submodule L of M is

<u>complemented</u> if there exists a submodule K of M such that L is a maximal submodule having the property that $K \cap L = 0$. In such a case, $K + L$ is an essential submodule of M.

<u>Lemma 4.1</u>. L is complemented in M if and only if $L = E(L) \cap M$ (i.e. L has no essential extension within M).

<u>Proof</u>. Suppose L is maximal such that $L \cap K = 0$. There is an isomorphism $E(M) \cong E(K) \oplus E(L)$ which reduces to the identity map on $K + L$. Therefore $E(L) \cap M \supset L$ and $E(L) \cap M \cap K = 0$, and by maximality of L it follows that $E(L) \cap M = L$.

Suppose conversely that L is "essentially closed" in M. By Zorn's lemma there exists a maximal submodule K of M such that $K \cap L = 0$. Then also L is maximal with respect to $L \cap K = 0$, for if $L' \supset L$ and $L' \cap K = 0$, then $L + K$ is an essential submodule of $L' + K$, which implies that L is essential in L' and hence $L = L'$.

It should be remarked here that $E(L)$ has in general no unique imbedding in $E(M)$.

<u>Proposition 4.2</u>. The following properties of a hereditary torsion theory are equivalent:

(a) The theory is stable.

(b) The torsion submodule $t(M)$ is complemented in M, for every module M.

(c) Every injective module is the direct sum of a torsion module and a torsion-free module.

Proof. (a) $\Rightarrow$ (b): Since $t(M)$ is a torsion module, we have $E(t(M)) \subset t(E(M))$ by stability. But $t(M)$ is an essential submodule of $t(E(M))$, for if $0 \neq L \subset t(E(M))$, then L is a torsion module and $L \cap M \neq 0$, so $L \cap t(M) \neq 0$. Therefore we must have $E(t(M)) = t(E(M))$. Left exactness of t now implies $t(M) = t(E(M)) \cap M = E(t(M)) \cap M$, and Lemma 4.1 gives (b).

(b) $\Rightarrow$ (c): If E is an injective module, then $t(E)$ is also injective by Lemma 4.1, so $t(E)$ splits off.

(c) $\Rightarrow$ (a): If M is a torsion module, then $E(M) = T \oplus F$ where T is a torsion module and F is torsion-free. Since $T \supset M$, we must have $F = 0$.

Proposition 4.3. Let r be a left exact preradical such that the torsion theory associated to $\bar{r}$ is stable. Then $\bar{r}(M)$ is the unique maximal essential extension of $r(M)$ in M, for every module M.

Proof. We first show that $r(M)$ is an essential submodule of $\bar{r}(M)$. Suppose $L \subset \bar{r}(M)$. If $L \cap r(M) = 0$, then $r(L) = 0$ by left exactness. This implies $0 = \bar{r}(L) = L$.

Suppose $r(M)$ is essential in some submodule L of M. By 4.2 there exists $K \subset L$ such that $\bar{r}(L)$ is maximal with respect to $\bar{r}(L) \cap K = 0$. But then $r(K) = 0$ so we must have $K \cap r(M) = 0$. Hence $K = 0$ and $\bar{r}(L) = L$ by the maximality condition. This gives $L \cap \bar{r}(M) = \bar{r}(L) = L$, and so $L \subset \bar{r}(M)$.

An important example of stable torsion is given by:

Proposition 4.4. The Goldie torsion theory is stable.

Proof. If M is a torsion module, consider the exact sequence

$$0 \longrightarrow M \longrightarrow E(M) \longrightarrow E(M)/M \longrightarrow 0$$

where also $E(M)/M$ is a torsion module by definition of the Goldie theory. It follows that $E(M)$ is a torsion module.

Reference: Gabriel [31] (p. 426).

§ 5. Topologies for a commutative noetherian ring

Our purpose is to determine all topologies for a commutative noetherian ring. The first step in that direction is taken by determining all injective modules.

Proposition 5.1. If A is a right noetherian ring, then every direct limit of injective modules is injective.

Proof. Let (E_α) be a direct family of injective modules and let $f : I \to \varinjlim E_\alpha$ be a homomorphism from a right ideal I. Since I is finitely generated, there exists an index β such that $f(I)$ lies in the image of E_β in $\varinjlim E_\alpha$. Choose a finitely generated submodule L of E_β which maps onto $f(I)$. The kernel K of L $f(I)$ is then also finitely generated. Since K goes to zero in $\varinjlim E_\alpha$, there exists $\gamma \geq \beta$ such that K goes to zero already in E_γ. We may then factor f over the injective module E_γ, and it follows that f may be extended to A.

Diagram for 5.1:

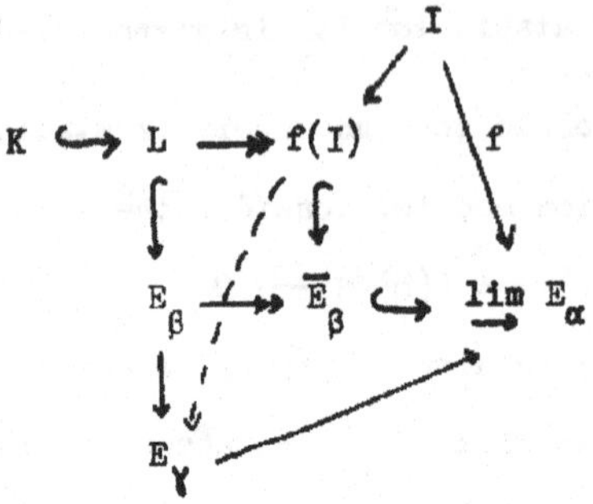

Corollary 5.2. If A is right noetherian, then every direct sum of injective modules is injective.

Proposition 5.3. If A is right noetherian, then every injective module is a direct sum of indecomposable injective modules.

Proof. Let E be an injective module. Consider all families (E_i) of indecomposable injective submodules such that the sum $\sum E_i$ in E is direct. By Zorn's lemma there exists a maximal such family $(E_i)_I$. From 5.2 we know that the sum $\bigoplus_I E_i$ is an injective module, so we can write $E = (\bigoplus_I E_i) \oplus E'$. We want to show that $E' = 0$, and for this it suffices to show that every injective module $E' \neq 0$ contains an indecomposable direct summand. Let x be any element $\neq 0$ in E'. Consider all injective submodules E'' of E such that $x \notin E''$. By Zorn's lemma and 5.1 there exists a maximal such submodule E''. We have $E' = E'' \oplus L$, and we assert that L is indecomposable. For if $L = L' \oplus L''$ with L' and $L'' \neq 0$, then $(E'' \oplus L') \cap (E'' \oplus L'') = E''$ so one of the modules $E'' \oplus L'$ or $E'' \oplus L''$ does not contain x, which contradicts the maximality of E''.

It is possible to prove the converse statement, namely that if every injective module is a direct sum of indecomposable modules, then A must be right noetherian (see e.g. Faith-Walker [30]).*)

Lemma 5.4. An injective module E is indecomposable if and only $E = E(A/I)$ where I is an irreducible right ideal.

Proof. Easy exercise.

For the rest of this § we assume A to be a commutative noetherian ring. We write $\mathrm{Spec}(A)$ for the family of prime ideals of A. For each module M we define

$$\mathrm{Ass}(M) = \{ \underline{p} \in \mathrm{Spec}(A) \mid \underline{p} = \mathrm{Ann}(x) \text{ for some } x \neq 0 \text{ in } M \}.$$

Lemma 5.5. If $M \neq 0$, then $\mathrm{Ass}(M) \neq \emptyset$.

Proof. The family of ideals $\mathrm{Ann}(y)$ for $y \neq 0$ has a maximal member $\mathrm{Ann}(x)$ since A is noetherian. We assert that $\mathrm{Ann}(x)$ is a prime ideal. Suppose $a, b \in A$ with $ab \in \mathrm{Ann}(x)$ but $b \notin \mathrm{Ann}(x)$. Then $bx \neq 0$ and since $\mathrm{Ann}(bx) \supset \mathrm{Ann}(x)$, we must have $\mathrm{Ann}(x) = \mathrm{Ann}(bx)$. Now $a \in \mathrm{Ann}(bx)$ implies $a \in \mathrm{Ann}(x)$, and $\mathrm{Ann}(x)$ is prime.

Proposition 5.6. There is a 1-1 correspondence between prime ideals of A and isomorphism classes of indecomposable injective A-modules, given by $\underline{p} \mapsto E(A/\underline{p})$.

---

*) Our "proof" of this result in a paper in Arkiv för Matematik, vol. 7 (Theorem 2 on p. 428) is insufficient; one has to use some kind of "counting" argument, as in [30].

Proof. If $\underline{p}$ is a prime ideal, then $\underline{p}$ is irreducible and $E(A/\underline{p})$ is indecomposable by 5.4. If $\underline{p}$ and $\underline{q}$ are prime ideals such that $E(A/\underline{p}) \cong E(A/\underline{q})$, we consider both $A/\underline{p}$ and $A/\underline{q}$ as submodules of $E(A/\underline{p})$, and we then have $A/\underline{p} \cap A/\underline{q} \neq 0$ since $E(A/\underline{p})$ is indecomposable. But $\mathrm{Ann}(x) = \underline{p}$ for every non-zero $x \in A/\underline{p}$ and likewise $\mathrm{Ann}(y) = \underline{q}$ for non-zero $y \in A/\underline{q}$, so it follows that $\underline{p} = \underline{q}$.
Finally, let E be an indecomposable injective module. By 5.5 there exists $\underline{p} \in \mathrm{Ass}(E)$, and since E is indecomposable we must have $E = E(A/\underline{p})$.

Let $\underline{F}$ be a topology on A, and let $(\underline{\underline{T}}, \underline{\underline{F}})$ be the corresponding torsion theory.

Proposition 5.7. $(\underline{\underline{T}}, \underline{\underline{F}})$ is generated by the cyclic modules of the form $A/\underline{p}$ where $\underline{p} \in \mathrm{Spec}(A) \cap \underline{F}$.

Proof. We must show that if a module M has the property that $\mathrm{Hom}(A/\underline{p}, M) = 0$ for all prime ideals $\underline{p}$ in $\underline{F}$, then M is torsion-free. We decompose the injective envelope of M as $E(M) = \bigoplus E(A/\underline{p})$. For each summand we have $A/\underline{p} \cap M \neq 0$, and for each non-zero $x \in A/\underline{p}$ we have $\mathrm{Ann}(x) = \underline{p}$ since $\underline{p}$ is prime. By the hypothesis on M this implies $\underline{p} \notin \underline{F}$, and hence each $A/\underline{p}$ is torsion-free. Then $E(M) = \bigoplus E(A/\underline{p})$ is torsion-free by heredity, and so M is torsion-free.

Lemma 5.8. For each finitely generated module M there exists a sequence of submodules $M = M_n \supset M_{n-1} \supset \ldots\ldots \supset M_o = (0)$ such that $M_i/M_{i-1} \cong A/\underline{p}_i$ where $\underline{p}_i \in \mathrm{Spec}(A)$.
M is a torsion module if and only if each $\underline{p}_i \in \underline{F}$.

Proof. Suppose we have constructed $M_o, \ldots, M_{i-1}$. If $M_{i-1} \neq M$, we have $\mathrm{Ass}(M/M_{i-1}) \neq \emptyset$ by 5.5, so there exists $M_i \subset M$ such that $M_i/M_{i-1} \cong A/\underline{p}_i$. Since M is noetherian, the chain will eventually stop. It is clear that M is a torsion module if and only if each M_i/M_{i-1} is a torsion module.

Theorem 5.9. A family $\underline{F}$ of ideals of A is a topology if and only if there exists $\underline{P} \subset \mathrm{Spec}(A)$ such that $\underline{F} = \{I \mid I \supset \underline{p}_o \cdot \ldots \cdot \underline{p}_n \text{ where } \underline{p}_i \in \underline{P}\}$.

Proof. Suppose $\underline{F}$ is a topology. If $I \in \underline{F}$, it follows from 5.8 that there exist $\underline{p}_o, \ldots, \underline{p}_n \in \mathrm{Spec}(A) \cap \underline{F}$ such that $\underline{p}_o \cdot \ldots \cdot \underline{p}_n$ annihilates A/I, i.e. $\underline{p}_o \cdot \ldots \cdot \underline{p}_n \subset I$. Conversely, if I contains a product of ideals in $\mathrm{Spec}(A) \cap \underline{F}$, then $I \in \underline{F}$ by § 3, Exercise 1.

On the other hand, if $\underline{P} \subset \mathrm{Spec}(A)$ is given and $\underline{F} = \{I \mid I \supset \underline{p}_o \cdot \ldots \cdot \underline{p}_n \text{ with } \underline{p}_i \in \underline{P}\}$, then $\underline{F}$ is clearly a pre-topology and is in fact a topology by § 3, Exercise 2.

Lemma 5.10. If $\underline{p} \in \mathrm{Spec}(A)$ and $x \in E(A/\underline{p})$, $x \neq 0$, then $\mathrm{Ann}(x)$ is an irreducible $\underline{p}$-primary ideal.

Proof. Since $E(A/\underline{p})$ is indecomposable, it is an injective envelope of $A/\mathrm{Ann}(x)$. Lemma 5.4 gives that $\mathrm{Ann}(x)$ is an irreducible ideal, hence primary ([83], p. 209). Let $\underline{q}$ be the prime ideal associated to $\mathrm{Ann}(x)$. Let n be the smallest integer such that $\underline{q}^n \subset \mathrm{Ann}(x)$. Considering $A/\underline{p}$ and $A/\mathrm{Ann}(x)$ as submodules of $E(A/\underline{p})$, we have $A/\underline{p} \cap \underline{q}^{n-1}/\mathrm{Ann}(x) \neq 0$. Hence there exists $a \in \underline{q}^{n-1}$ such that $\bar{a} \neq 0$ in $A/\mathrm{Ann}(x)$ and

$\text{Ann}(\bar{a}) = \underline{p}$. We have $\underline{q}a \subset \text{Ann}(x)$, so $\text{Ann}(\bar{a}) \supset \underline{q}$. But on the other hand, if $b\bar{a} = 0$, then $ba \in \text{Ann}(x)$ and $b \in \underline{q}$ since $\text{Ann}(x)$ is $\underline{q}$-primary. Hence $\underline{q} = \text{Ann}(\bar{a}) = \underline{p}$.

Corollary 5.11. If $x \in E(A/\underline{p})$, then $\underline{p}^n x = 0$ for some $n > 0$.
Proof. Since $\text{Ann}(x)$ is $\underline{p}$-primary, $\underline{p}^n \subset \text{Ann}(x)$ for some n.

Proposition 5.12. Every hereditary torsion theory on A-Mod is stable.
Proof. Let M be a torsion module. We decompose $E(M)$ as $E(M) = \oplus E(A/\underline{p}_i)$. Since each $A/\underline{p}_i$ has non-zero intersection with the torsion module M, we must have $\underline{p}_i \in \underline{F}$. It follows from 5.9 and 5.11 that each $E(A/\underline{p}_i)$ is a torsion module, and hence $E(M)$ is a torsion module.

Example:
Let A be a commutative noetherian ring and let s be the preradical assigning to each module M its socle $s(M)$ (§ 3, Example 5). From Proposition 4.3 we get that the corresponding torsion submodule $\bar{s}(M)$ is the maximal essential extension of $s(M)$ in M.

Exercises:
Let A be a commutative noetherian ring with a topology $\underline{F}$.
1. Show that a module M is a torsion module if and only if $\text{Ass}(M) \subset \underline{F}$.
2. Let M be a finitely generated module and let I be any ideal. Show that $I^n M = 0$ for some $n > 0$ if and only if

I is contained in every $\underline{p} \in \mathrm{Ass}(M)$.

References: Gabriel [31] (ch. 5, §4 and 5), Matlis [52], [53].

§ 6. $\underline{F}$-injective modules

Let $\underline{F}$ be a topology on A .

Definition. A module E is $\underline{F}$-injective if $\mathrm{Ext}^1(A/I,E) = 0$ for every $I \in \underline{F}$.

Before giving some alternative descriptions of $\underline{F}$-injectivity, we will introduce a convenient terminology. If M is a module, a submodule L of M is called an $\underline{F}$-submodule if $(L:x) \in \underline{F}$ for every $x \in M$, i.e. if M/L is a torsion module. The family of all $\underline{F}$-submodules of M is a filter (as a consequence of T3 and T4) which we denote by $\underline{F}(M)$. Note that T 1 implies $\underline{F}(A) = \underline{F}$.

Lemma 6.1. If $K \in \underline{F}(L)$ and $L \in \underline{F}(M)$, then $K \in \underline{F}(M)$.

Proof. We have an exact sequence $0 \to L/K \to M/K \to M/L \to 0$ where L/K and M/L are torsion modules. Then also M/K is torsion.

Proposition 6.2. The following properties of a module E are equivalent:

(a) E is $\underline{F}$-injective.

(b) $\mathrm{Ext}^1(M,E) = 0$ for every torsion module M .

(c) If $L \in \underline{F}(M)$, then every homomorphism $L \to E$ may be extended to a homomorphism $M \to E$.

Proof. (a) ⇒ (c): Let $L \in \underline{F}(M)$ and $f:L \to E$ be given. In the usual way we may assume that there is a maximal extension $f':L' \to E$ of f , where $L \subset L' \subset M$. Then also $L' \in \underline{F}(M)$ by T 3. Suppose there exists $x \in M$ such that $x \notin L'$. Let I =

$= (L':x) \in \underline{F}$, and let $\alpha : I \to E$ be the homomorphism $\alpha(a) = f'(xa)$. By (a) we may extend α to A , i.e. there exists $y \in E$ such that $\alpha(a) = ya$. We may then define $g : L' + xA \to E$ as $g(z + xa) = f'(z) + ya$. g extends f' , which is a contradiction.

(c) $\Rightarrow$ (b): We choose an exact sequence $0 \to K \to F \to M \to 0$ where F is free. Since $\mathrm{Ann}(x) \in \underline{F}$ for every $x \in M$, we have $K \in \underline{F}(M)$. The exact sequence

$$\mathrm{Hom}(F,E) \longrightarrow \mathrm{Hom}(K,E) \longrightarrow \mathrm{Ext}^1(M,E) \longrightarrow 0 \quad ,$$

together with (c), shows that $\mathrm{Ext}^1(M,E) = 0$.

(b) $\Rightarrow$ (a) is trivial.

Definition. An $\underline{F}$-injective envelope of M is a monomorphism $M \hookrightarrow E$ such that E is $\underline{F}$-injective and $M \in \underline{F}(E)$.

Proposition 6.3. The submodule $\{x \in E(M) \mid (M:x) \in \underline{F}\}$ of $E(M)$ is an $\underline{F}$-injective envelope of M .

Proof. Clearly the subset in question is a submodule of $E(M)$. It suffices to show that $E' = \{x \in E(M) \mid (M:x) \in \underline{F}\}$ is $\underline{F}$-injective. Suppose we are given $f : I \to E'$ with $I \in \underline{F}$. f may in any case be extended to a homomorphism $g : A \to E(M)$, so we get a commutative diagram

$$\begin{array}{ccccccccc}
0 & \longrightarrow & I & \longrightarrow & A & \longrightarrow & A/I & \longrightarrow & 0 \\
 & & \downarrow f & & \downarrow g & & \downarrow h & & \\
0 & \longrightarrow & E' & \longrightarrow & E(M) & \longrightarrow & E(M)/E' & \longrightarrow & 0 \\
 & & \downarrow & & \downarrow & & \| & & \\
0 & \longrightarrow & E'/M & \longrightarrow & E(M)/M & \longrightarrow & E(M)/E' & \longrightarrow & 0
\end{array}$$

If $h \neq 0$, then $E(M)/E'$ contains a non-zero torsion submodule, and from the lowest row we see that this contradicts the fact that

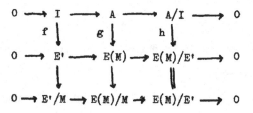

E' is the maximal submodule of E(M) containing M as an $\underline{F}$-submodule. Hence $h = 0$, and g actually maps A into E' .

It follows from this result that the $\underline{F}$-injective envelope of M is unique up to isomorphism; it will be denoted as $E_{\underline{F}}(M)$. We will give another useful description of the $\underline{F}$-injective envelope. Recall first that a hereditary torsion theory is cogenerated by some injective module (§ 2, Exercise 4).

<u>Proposition 6.4</u>. If C is an injective module cogenerating the torsion theory associated to $\underline{F}$, then

$E_{\underline{F}}(M) = \{ x \in E(M) \mid f(x) = 0$ for all $f: E(M) \to C$ with $f(M) = 0 \}$

for every module M .

<u>Proof</u>. Suppose $x \in E_{\underline{F}}(M)$ and consider any $f: E(M) \to C$ with $f(M) = 0$. There exists $I \in \underline{F}$ such that $xI \subset M$. Then $f(x)I = f(xI) = 0$ and $f(x) = 0$ since C is torsion-free.

Conversely, assume x has the property that $f(x) = 0$ for $f: E(M) \to C$ with $f(M) = 0$. We want to show that the right ideal $I = (M:x)$ is in $\underline{F}$, i.e. to show that $\mathrm{Hom}(A/I, C) = 0$. If $g: A/I \to C$ is given, we may extend it to give a commutative diagram

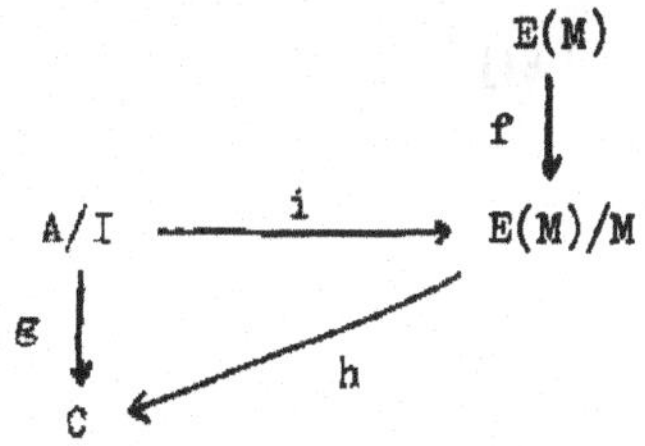

where $i(\bar{a}) = xa$. Since $hf(M) = 0$, we have $hf(x) = 0$ which implies $g = 0$.

Examples:

1. Proposition 6.2 holds also when $\underline{F}$ is only a pretopology. If e.g. $\underline{F}$ is the family of essential right ideals, then the filter $\underline{F}(M)$ consists of the essential submodules of M, while the $\underline{F}$-injective modules are precisely the injective modules.

2. Let $\underline{D}$ be the family of dense right ideals. The members of $\underline{D}(M)$ are usually called dense submodules of M.

Exercises:

1. An exact sequence $0 \to L \to M \to N \to 0$ is called $\underline{F}$-pure if for every $x \in N$ with $\mathrm{Ann}(x) \in \underline{F}$ there exists $y \in M$ mapping onto x such that $\mathrm{Ann}(y) = \mathrm{Ann}(x)$. L is then an $\underline{F}$-pure submodule of M. Show that the following properties of a module L are equivalent:

(i) L is $\underline{F}$-injective.

(ii) Every exact sequence $0 \to L \to M \to N \to 0$ is $\underline{F}$-pure.

(iii) L is an $\underline{F}$-pure submodule of $E(L)$.

References: Walker and Walker [81].

Chapter 2. Categories of modules of quotients

§ 7. Construction of rings and modules of quotients

Let $\underline{F}$ be a right additive topology on the ring A. For each right module M we will define its module of quotients with respect to $\underline{F}$. This we will do in two steps. The first step we take is to define

$$M_{(\underline{F})} = \varinjlim \operatorname{Hom}_A(I,M) \quad , \qquad I \in \underline{F} ,$$

where the direct limit is taken over the downwards directed family $\underline{F}$ of right ideals. We want to give $A_{(\underline{F})}$ the structure of a ring and $M_{(\underline{F})}$ that of a right $A_{(\underline{F})}$-module. For this we need:

Lemma 7.1. If I , $J \in \underline{F}$ and $\alpha : I \to A$ is a homomorphism, then $\alpha^{-1}(J) \in \underline{F}$.

Proof. For each $a \in I$ we have $(\alpha^{-1}(J):a) = \{ b \mid \alpha(ab) \in J \} = (J:\alpha(a)) \in \underline{F}$ by T 1, so $\alpha^{-1}(J) \in \underline{F}$ by T 2.

We define a pairing $M_{(\underline{F})} \times A_{(\underline{F})} \longrightarrow M_{(\underline{F})}$ as follows: suppose $x \in M_{(\underline{F})}$, $a \in A_{(\underline{F})}$ are represented by $\xi : J \to M$ and $\alpha : I \to A$; we then define $xa \in M_{(\underline{F})}$ to be represented by the composed map

$$\alpha^{-1}(J) \longrightarrow J \longrightarrow M \quad ,$$

using Lemma 7.1. It is easy to see that xa is well-defined, i.e. is independent of the choice of the representing homomorphisms ξ , α . One also easily verifies that the pairing $M_{(\underline{F})} \times A_{(\underline{F})} \longrightarrow M_{(\underline{F})}$ is biadditive. When $M = A$, this makes $A_{(\underline{F})}$ into a ring, and in the general case it makes $M_{(\underline{F})}$ into an $A_{(\underline{F})}$-module.

There are canonical homomorphisms

$$\varphi_M : M \cong \mathrm{Hom}_A(A,M) \longrightarrow \varinjlim \mathrm{Hom}_A(I,M) = M_{(\underline{F})} \quad .$$

In particular, φ_A is a ring homomorphism. By pullback along φ_A we may consider each $M_{(\underline{F})}$ as an A-module. φ_M is then A-linear. The assignment $M \mapsto M_{(\underline{F})}$ is a functor $\text{Mod-}A \to \text{Mod-}A_{(\underline{F})}$ which is left exact, because Hom is left exact and direct limits are exact in Mod-A . We denote by $L:\text{Mod-}A \to \text{Mod-}A$ this functor followed by the forgetful functor $\text{Mod-}A_{(\underline{F})} \to \text{Mod-}A$.

Let t denote the torsion radical associated to $\underline{F}$.

<u>Lemma 7.2.</u> $\mathrm{Ker}(M \xrightarrow{\varphi_M} M_{(\underline{F})}) = t(M)$.

<u>Proof.</u> If $\varphi_M(x) = 0$, then x goes to zero already in some $\mathrm{Hom}(I,M)$, $I \in \underline{F}$. This means that the map $a \mapsto xa$ is zero on I , i.e. $xI = 0$ and $x \in t(M)$. This argument may be reversed.

<u>Lemma 7.3.</u> M is a torsion module if and only if $M_{(\underline{F})} = 0$.

<u>Proof.</u> If $M_{(\underline{F})} = 0$, then $t(M) = \mathrm{Ker}\,\varphi_M = M$ by 7.2. Suppose on the other hand that M is a torsion module. Let $x \in M_{(\underline{F})}$ be represented by $\xi : J \to M$. If we can show that $\mathrm{Ker}\,\xi \in \underline{F}$, it will follow that $x = 0$. For each $a \in J$ there exists $I_a \in \underline{F}$ such that $\xi(a)I_a = 0$. Put $K = \sum_{a \in J} aI_a$. Then $K \subset \mathrm{Ker}\,\xi$, and for each $a \in J$ we have $(K:a) \supset I_a$, so $(K:a) \in \underline{F}$ and it follows that $K \in \underline{F}$ by T 2. Hence $\mathrm{Ker}\,\xi \in \underline{F}$.

<u>Lemma 7.4.</u> If $x \in M_{(\underline{F})}$ is represented by $\xi : I \to M$, $I \in \underline{F}$, then the diagram

$$\begin{array}{ccc} I & \hookrightarrow & A \\ {\scriptstyle\xi}\downarrow & & \downarrow{\scriptstyle\beta} \\ M & \xrightarrow[\varphi_M]{} & M_{(\underline{F})} \end{array}$$

commutes, where $\beta(a) = xa$.

Proof. If $a \in I$, then xa is represented by the composed morphism $A = (I:a) \to I \to M$ given by $b \mapsto \xi(ab) = \xi(a)b$, so $xa = \varphi_M \xi(a)$.

An immediate consequence of this is:

Lemma 7.5. $\operatorname{Coker} \varphi_M$ is a torsion module.

As the second step in our construction of modules of quotients we apply the functor L once more to $M_{(\underline{F})}$. This results in a ring $A_{\underline{F}}$ (called the ring of quotients of A with respect to $\underline{F}$) and an $A_{\underline{F}}$-module $M_{\underline{F}}$ for each A-module M. The description of this is considerably simplified by:

Lemma 7.6. L carries the monomorphism $M/t(M) \hookrightarrow M_{(\underline{F})}$ into an isomorphism $(M/t(M))_{(\underline{F})} \xrightarrow{\approx} M_{\underline{F}}$.

Proof. Apply the left exact functor L to the exact sequence

$$0 \longrightarrow M/t(M) \longrightarrow M_{(\underline{F})} \longrightarrow \operatorname{Coker} \varphi_M \longrightarrow 0$$

and apply Lemmas 7.5 and 7.3.

We have thus obtained the formula

$$M_{\underline{F}} = \varinjlim \operatorname{Hom}_A(I, M/t(M)) \qquad , \quad I \in \underline{F} .$$

One verifies that the ring structure of $A_{\underline{F}}$ and the module structure of $M_{\underline{F}}$ are given by the following pairing $M_{\underline{F}} \times A_{\underline{F}} \to M_{\underline{F}}$:

let $x \in M_{\underline{F}}$ be represented by $\xi: J \to M/t(M)$,

$a \in A_{\underline{F}}$ — " — $\alpha: I \to A/t(A)$;

ξ induces $J/t(J) \to M/t(M)$ and we have $J/t(J) \hookrightarrow A/t(A)$ by left exactness of t; $xa \in M_{\underline{F}}$ is represented by

$$\alpha^{-1}(J/t(J)) \longrightarrow J/t(J) \longrightarrow M/t(M) .$$

For each $f: M \to N$ in Mod-A one gets $f_{\underline{F}}: M_{\underline{F}} \to N_{\underline{F}}$ in Mod-$A_{\underline{F}}$, which gives a functor $q:$ Mod-$A \to$ Mod-$A_{\underline{F}}$. There are canonical homomorphisms $\psi_M: M \to M_{\underline{F}}$ of A-modules; in particular $\psi_A: A \to A_{\underline{F}}$ is a ring homomorphism. For each $f \in \mathrm{Hom}_A(M,N)$ we get a commutative diagram

$$\begin{array}{ccc} M & \xrightarrow{\ f\ } & N \\ {\scriptstyle \psi_M}\downarrow & & \downarrow{\scriptstyle \psi_N} \\ M_{\underline{F}} & \xrightarrow[f_{\underline{F}}]{} & N_{\underline{F}} \end{array}$$

Note that $\mathrm{Ker}\,\psi_M = t(M)$ and that $\mathrm{Coker}\,\psi_M$ also is a torsion module.

When the torsion theory is stable (§ 4), the formula for $M_{\underline{F}}$ simplifies somewhat:

<u>Proposition 7.7.</u> When the torsion theory is stable, one has $M_{\underline{F}} = \varinjlim \mathrm{Hom}_A(I,M)$ for every module M.

<u>Proof.</u> Since $\varinjlim$ is an exact functor, the sequence

$$0 \longrightarrow t(M) \longrightarrow M \longrightarrow M/t(M) \longrightarrow 0$$

induces an exact sequence

$$0 \to \varinjlim \mathrm{Hom}(I,t(M)) \longrightarrow \varinjlim \mathrm{Hom}(I,M) \longrightarrow \varinjlim \mathrm{Hom}(I,M/t(M)) \longrightarrow$$
$$\longrightarrow \varinjlim \mathrm{Ext}^1(I,t(M)) .$$

The first term is zero by Lemma 7.3. If E is an injective envelope of $t(M)$, then E is a torsion module by hypothesis. The sequence

$$0 \longrightarrow t(M) \longrightarrow E \longrightarrow E/t(M) \longrightarrow 0$$

induces the exact sequence

$$\varinjlim \mathrm{Hom}(I,E/t(M)) \longrightarrow \varinjlim \mathrm{Ext}^1(I,t(M)) \longrightarrow 0$$

where the first term is zero, again by Lemma 7.3. Hence the last term of the long exact sequence is zero.

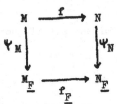

We want to study the image category of the functor q. For this purpose we introduce:

__Definition.__ M_A is $\underline{F}$-__closed__ if the canonical maps

$$M \cong \mathrm{Hom}_A(A,M) \longrightarrow \mathrm{Hom}_A(I,M)$$

are isomorphisms for all $I \in \underline{F}$.

Thus M is $\underline{F}$-closed if and only if M is both torsion-free and $\underline{F}$-injective (as defined in § 6). For every $\underline{F}$-closed module M we get an isomorphism $\psi_M : M \xrightarrow{\cong} M_{\underline{F}}$. Conversely we have:

__Proposition 7.8.__ $M_{\underline{F}}$ is $\underline{F}$-closed for every module M_A.

__Proof.__ To show that $M_{\underline{F}}$ is torsion-free, it suffices to show that if M is torsion-free, then $M_{(\underline{F})}$ is torsion-free. Suppose $x \in M_{(\underline{F})}$ and $xJ = 0$ for some $J \in \underline{F}$. Let x be represented by $\xi : I \to M$. By Lemma 7.4 we have a commutative diagram

$$\begin{array}{ccc} I & \hookrightarrow & A \\ {\scriptstyle \xi}\downarrow & & \downarrow{\scriptstyle \beta} \\ M & \xrightarrow[\varphi_M]{} & M_{(\underline{F})} \end{array} \qquad \beta(a) = xa$$

so $\varphi_M \xi$ is zero when restricted to $I \cap J \in \underline{F}$. But φ_M is a monomorphism, so $\xi | I \cap J = 0$ and $x = 0$.

Next we show that $M_{\underline{F}}$ is $\underline{F}$-injective. Suppose we are given $f : I \to M_{\underline{F}}$ with $I \in \underline{F}$. Consider the pullback diagram

$$\begin{array}{ccc} J & \hookrightarrow & I \\ {\scriptstyle g}\downarrow & & \downarrow{\scriptstyle f} \\ M/t(M) & \overset{}{\underset{\bar{\psi}}{\hookrightarrow}} & M_{\underline{F}} \end{array}$$

where also J is a right ideal. We have $I/J \cong \mathrm{Coker}\,\bar{\psi} =$

= Coker ψ_M , which is a torsion module, so it follows from T 2 that $J \in \underline{F}$. Lemma 7.4 now tells us that we may extend g to a homomorphism $h: A \to M_{\underline{F}}$. h is also an extension of f , because $h|I$ and f are equal on J , and therefore their difference factors over the torsion module I/J , and $M_{\underline{F}}$ torsion-free then implies $h|I = f$.

Corollary 7.9. The full subcategory of Mod-$A_{\underline{F}}$ consisting of modules of the form $M_{\underline{F}}$ is equivalent to the full subcategory of Mod-A consisting of $\underline{F}$-closed modules.

Let $\underline{C}$ be the full subcategory of Mod-A consisting of $\underline{F}$-closed modules. We have a number of interesting functors:

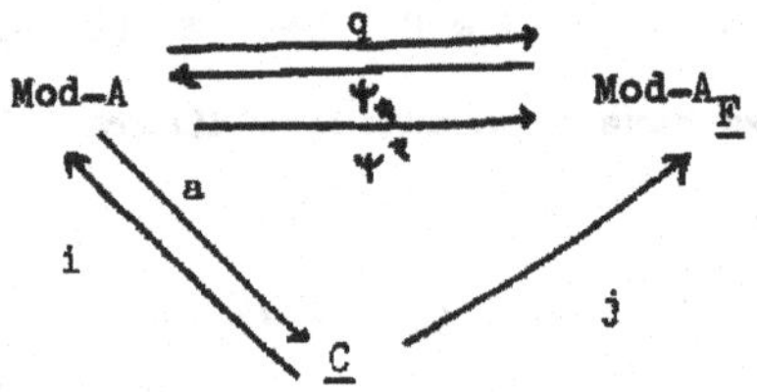

ψ_* is the forgetful functor,

q is the functor $M \mapsto M_{\underline{F}}$,

ψ^* is the functor $M \mapsto M \otimes_A A_{\underline{F}}$,

i is the inclusion functor,

a is the functor $M \mapsto M_{\underline{F}}$

j considers each $\underline{F}$-closed module as an $A_{\underline{F}}$-module and is full and faithful.

We have $ja = q$ and $i = \psi_* j$. ψ induces a natural equivalence $ai \simeq Id_{\underline{C}}$. On the other hand, $ia = LL = \psi_* q$.

<u>Proposition 7.10.</u> a is a left adjoint of i .

<u>Proof.</u> We must show that the canonical map

$$\mathrm{Hom}_A(M_{\underline{F}},N) \longrightarrow \mathrm{Hom}_A(M,N)$$

is an isomorphism when N is $\underline{F}$-closed.

$$\begin{array}{ccc} M & \xrightarrow{\ \Psi_M\ } & M_{\underline{F}} \\ {\scriptstyle f}\downarrow & & \downarrow{\scriptstyle f_{\underline{F}}} \\ N & \xrightarrow[\ \cong\ ]{} & N_{\underline{F}} \end{array}$$

It clearly is an epimorphism, so it remains to show that it is a monomorphism. The exact sequence

$$0 \longrightarrow M/t(M) \longrightarrow M_{\underline{F}} \longrightarrow \mathrm{Coker}\,\Psi_M \longrightarrow 0$$

induces

$$\mathrm{Hom}(\mathrm{Coker}\,\Psi_M,N) \longrightarrow \mathrm{Hom}(M_{\underline{F}},N) \longrightarrow \mathrm{Hom}(M/t(M),N)$$

where the first term is zero since $\mathrm{Coker}\,\Psi_M$ is a torsion module. The desired conclusion now follows from the observation that $\mathrm{Hom}(M/t(M),N) \cong \mathrm{Hom}(M,N)$.

We will show in the following sections that the category $\underline{C}$ of $\underline{F}$-closed modules is very well-behaved, in fact it is an abelian category with exact direct limits, although the inclusion functor i is not exact.

<u>Example:</u> Let $\underline{D}$ be the family of dense right ideals. The ring $A_{\underline{D}}$ is called the <u>maximal</u> (or <u>complete</u>) ring of quotients of A and will be denoted by Q_m . Since A is $\underline{D}$-torsion-free, we have $Q_m = \varinjlim \mathrm{Hom}(I,A)$, $I \in \underline{D}$.

Exercises:

1. Show that if $\underline{E} \subset \underline{F}$ are topologies, there is a ring homomorphism $A_{\underline{E}} \to A_{\underline{F}}$.
2. Show that if A is a commutative ring, then also every $A_{\underline{F}}$ is commutative. (Hint: one can reduce the problem to showing that if $I \in \underline{F}$ and $\alpha, \beta : I \to A$, then $\alpha\beta(x) = \beta\alpha(x)$ for $x \in I^2$).
3. Let $A = K[X,Y]$ where K is a field, and let $\underline{m}$ be the maximal ideal $\underline{m} = (X,Y)$. Consider the topology $\underline{F} = \{ I \mid I \supset \underline{m}^n$ for some $n\}$. Show that:

   (i) $A_{\underline{F}} = A$.

   (ii) If $M = A/(X)$, then $M_{\underline{F}} = K[\bar{Y}, 1/\bar{Y}]$, where $\bar{Y}$ is the class of Y in M. (So if $f: A \to M$ is the canonical epimorphism, then $f_{\underline{F}}$ is not an epimorphism, and the functor q is not exact).
4. Let $\underline{F}$ be a topology on A and assume A $\underline{F}$-torsion-free (so A is a subring of $A_{\underline{F}}$). For each right A-submodule I of $A_{\underline{F}}$, put $I^* = \{ q \in A_{\underline{F}} \mid qI \subset A\}$.

   (i) If $I \in \underline{F}(A_{\underline{F}})$ (cf. § 6), show that $I^* \cong \mathrm{Hom}_A(I,A)$.

   (ii) Call I $\underline{F}$-invertible if there exist $a_1, \ldots, a_n \in I$ and $q_1, \ldots, q_n \in I^*$ such that $1 = \sum a_i q_i$. Show that the following properties of I are equivalent:

   (a) I is $\underline{F}$-invertible.

   (b) $II^* = \{ q \in A_{\underline{F}} \mid qI \subset I \}$.

   (c) I is a finitely generated projective module and $I \in \underline{F}(A_{\underline{F}})$.

   (Hint: cf. [13], p. 132).

5. Show that for any ring A and module M_A one has

$$E(M)/Z(M) = \varinjlim \mathrm{Hom}(I,M)$$

where I runs through the downwards directed family of essential right ideals of A.

References: Bourbaki [10], (p. 157 and following), Gabriel [31], (p. 411 and following), Goldman [33], Roos [66] (ch. 1).

§ 8. Modules of quotients and $\underline{F}$-injective envelopes

Let $\underline{F}$ be a topology on A.

Proposition 8.1. If M is a torsion-free module, then $M_{\underline{F}} \cong E_{\underline{F}}(M)$ as $A_{\underline{F}}$-modules.

Proof. $M_{\underline{F}}$ is $\underline{F}$-injective and $M_{\underline{F}}/M$ is a torsion module by 7.5, so $M_{\underline{F}}$ is an $\underline{F}$-injective envelope of M. If $E_{\underline{F}}(M) = \{x \in E(M) \mid (M:x) \in \underline{F}\}$ for a fixed injective envelope $E(M)$ of M, then the isomorphism $M_{\underline{F}} \cong E_{\underline{F}}(M)$ is $A_{\underline{F}}$-linear by 7.9.

Let us desribe explicitly the $A_{\underline{F}}$-module structure of $E_{\underline{F}}(M)$ for a torsion-free module M. Suppose $x \in E_{\underline{F}}(M)$, and $q \in A_{\underline{F}}$ is represented by $\alpha: I \to A/t(A)$. Since $E_{\underline{F}}(M)$ is $\underline{F}$-closed, there exists a unique $y \in E_{\underline{F}}(M)$ such that $x\alpha(a) = ya$ for all $a \in I$ (note that $E_{\underline{F}}(M)$ is a module over $A/t(A)$), and then $xq = y$. This description of the module structure is applicable also for the $\underline{F}$-closed module $E(M)$.

Proposition 8.2. If M is torsion-free, then $E(M)$ is an injective envelope of $M_{\underline{F}}$ in Mod-$A_{\underline{F}}$.

Proof. The inclusion map $M_{\underline{F}} \cong E_{\underline{F}}(M) \longrightarrow E(M)$ is $A_{\underline{F}}$-linear by 7.8. Since it is an essential monomorphism in Mod-A, it is obviously essential also in Mod-$A_{\underline{F}}$. It only remains to show that $E(M)$ is injective as an $A_{\underline{F}}$-module. This follows from:

Lemma 8.3. Every torsion-free injective A-module is injective over $A_{\underline{F}}$.

Proof. Let E be torsion-free injective over A. Suppose $N' \longrightarrow N$ is any monomorphism in Mod-$A_{\underline{F}}$ and $f: N' \longrightarrow E$ is $A_{\underline{F}}$-linear. f extends to a homomorphism $g: N \longrightarrow E$ in Mod-A. For each $x \in N$, consider the two maps $A_{\underline{F}} \longrightarrow E$ given by $g'(q) = g(xq)$ and $g''(q) = g(x)q$. They are both A-linear and coincide on A. Hence $g'-g''$ factors over a homomorphism Coker $\psi_A \longrightarrow E$. But Coker ψ_A is a torsion module (Lemma 7.5) and E is torsion-free, so $g' = g''$, and g is $A_{\underline{F}}$-linear.

For the remaining part of this § we will assume that A_A is torsion-free. Thus $\underline{F}$ is contained in the the topology $\underline{D}$ of dense right ideals, and the torsion theory is cogenerated by an injective module $C = E(A) \oplus F$. Let H be the endomorphism ring of the module C_A, and consider C as bimodule ${}_HC_A$. The ring $\mathrm{Hom}_H(C,C)$ is usually called the double centralizer of C. Since C is $\underline{F}$-closed, H is also the endomorphism ring of C as an $A_{\underline{F}}$-module. It follows that there is a commutative diagram of canonical ring homomorphisms

$$\begin{array}{ccc} A & \searrow & \\ \downarrow & & \mathrm{Hom}_H(C,C) \\ A_{\underline{F}} & \nearrow_{\lambda} & \end{array}$$

Theorem 8.4. λ is an isomorphism between $A_{\underline{F}}$ and the double centralizer of the cogenerating injective C .

Proof. We will exhibit an inverse μ of λ . Suppose $\beta \in \mathrm{Hom}_H(C,C)$. Let $p: C \to E(A)$ be the canonical projection. We want to show that $p\beta(1) \in E_{\underline{F}}(A)$ by using Proposition 6.4. Let $f: E(A) \to C$ be any A-linear map such that $f(1) = 0$. Extend f to $\bar{f}: C \to C$ by defining $\bar{f}(F) = 0$, where $C = E(A) \oplus F$. Then $f(p\beta(1)) = \bar{f}(\beta(1)) = \beta(f(1)) = 0$ since β is H-linear. Hence $p\beta(1) \in E_{\underline{F}}(A) \cong A_{\underline{F}}$. We may now define μ as $\mu(\beta) = p\beta(1) \in A_{\underline{F}}$.

μ is an additive map. Clearly $\mu\lambda = \mathrm{id.}$. Then $\mu\lambda\mu = \mu$, and it only remains to show that μ is a monomorphism . Thus we must show that if $\beta(1) \in F$, then $\beta = 0$. For every $x \in C$ there exists $h \in H$ such that $h(1) = x$ and $h(F) = 0$. Then $\beta(x) = \beta(h(1)) = h(\beta(1)) = 0$.

Examples:

1. Let $\underline{F}$ be the Goldie topology (§ 3, Example 1). For each torsion-free (i.e. non-singular) module we get $M_{\underline{F}} = E(M)$.
2. The maximal right ring of quotients of A is the double centralizer of $E(A)$.

Exercises:

Let E be an injective module. The E-dominant dimension of M_A is said to be $\geqslant n$ (notation: $E\text{-dom.dim } M \geqslant n$) if there exists an exact sequence

$$0 \to M \to E_1 \to \ldots\ldots\ldots \to E_n$$

where each E_i is a direct product of copies of E . Consider the hereditary torsion theory cogenerated by E . Show that:

(i) M is torsion-free if and only if E-dom.dim $M \geq 1$.

(ii) M is closed if and only if E-dom.dim $M \geq 2$. (Hint: M is $\underline{F}$-injective if and only if $E(M)/M$ is $\underline{F}$-torsion-free).

References: Lambek [47], [48], [49], Morita [57], Tachikawa [74], Turnidge [79], Wong and Johnson [114].

§ 9. Coreflective subcategories of Mod-A

We prepare the study of the category of $\underline{F}$-closed modules by a consideration of a more general situation:

Definition. A full subcategory $\underline{C}$ of Mod-A is called coreflective if the inclusion functor $i : \underline{C} \to$ Mod-A has a left adjoint a .

In such a case there exists a natural transformation $\psi : 1 \to ia$ such that the bijection

$$\mathrm{Hom}_A(a(M), N) \longrightarrow \mathrm{Hom}_A(M, N)$$

where $N \in \underline{C}$, is given by $\alpha \mapsto \alpha\psi_M$.

Examples:

1. If $\underline{F}$ is a toplogy on A , then the category of $\underline{F}$-closed modules is coreflective in Mod-A .
2. If $(\underline{\underline{T}}, \underline{\underline{F}})$ is a torsion theory for Mod-A , then the category of torsion-free modules is coreflective in Mod-A , with $a(M) = M/t(M)$.

Let $\underline{C}$ be coreflective in Mod-A . If M is a module in $\underline{C}$, we may choose $a(M) = M$ with the identity map as Ψ_M .

Lemma 9.1. If there exists $\alpha : a(M) \to M$ such that $\alpha \Psi_M = 1_M$, then Ψ_M is an isomorphism.

Proof. From the preceding remark it follows that $\Psi_{a(M)}$ is the identity map. The morphism α induces by naturality of ψ a commutative diagram

$$
\begin{array}{ccc}
a(M) & \xrightarrow{\ \alpha\ } & M \\
{\scriptstyle 1 = \Psi_{a(M)}}\big\downarrow & & \big\downarrow{\scriptstyle \Psi_M} \\
a^2(M) & \xrightarrow[a(\alpha)=1]{} & a(M)
\end{array}
$$

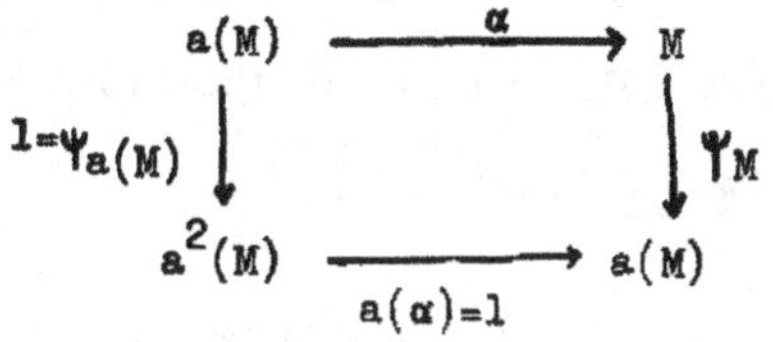

Hence $\Psi_M \alpha = a(\alpha) \Psi_{a(M)} = 1_{a(M)}$, and so α is the invers of Ψ_M .

Proposition 9.2. A coreflective subcategory of Mod-A has arbitrary limits and colimits.

Proof. Let $\underline{D}$ be a small category and $G : \underline{D} \to \underline{C}$ a functor. Then $\varprojlim iG$ exists and we denote it by M , and let $\alpha_d : M \to G(d)$ denote the canonical projections. Since $G(d) \varepsilon \underline{C}$, there exist $\beta_d : a(M) \to G(d)$ such that $\beta_d \Psi_M = \alpha_d$.

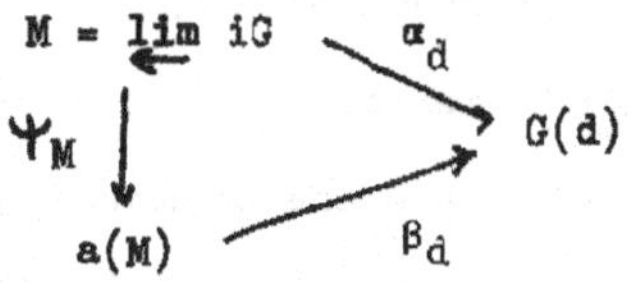

$$
\begin{array}{ccc}
M = \varprojlim iG & \xrightarrow{\ \alpha_d\ } & \\
{\scriptstyle \Psi_M}\big\downarrow & \searrow & G(d) \\
a(M) & \xrightarrow[\beta_d]{} &
\end{array}
$$

The family $\{\beta_d\}_{d \varepsilon \underline{D}}$ is compatible with the morphisms in $\underline{D}$, for if $\lambda : d \to d'$ in $\underline{D}$, then $G(\lambda)\beta_d \Psi_M = G(\lambda)\alpha_d = \alpha_{d'} = \beta_{d'} \Psi_M$ and hence $G(\lambda)\beta_d = \beta_{d'}$. It is therefore induced a map $\beta: a(M) \to M$ such that $\alpha_d\beta = \beta_d$ for all $d \in \underline{D}$. Then $\alpha_d\beta\Psi_M = \beta_d\Psi_M = \alpha_d$, so $\beta\Psi_M = 1_M$. From 9.1 it follows that Ψ_M is an isomorphism, and it is then clear that $a(M)$ is a limit for G in $\underline{C}$. Note that we have obtained the formula

$$i(\varprojlim G) = \varprojlim iG$$

which also follows from the fact that a right adjoint functor always commutes with limits (when these exist).

To prove that $\varinjlim G$ exists in $\underline{C}$ is easier, because the left adjoint a preserves colimits and we therefore have $a(\varinjlim iG) = \varinjlim aiG = \varinjlim G$, since $ai \cong 1$.

In other words, limits in $\underline{C}$ may be computed in Mod-A , while colimits in $\underline{C}$ are taken in Mod-A and then coreflected into $\underline{C}$.

Proposition 9.3. If $\underline{C}$ is coreflective in Mod-A and $a:\text{Mod-A} \to \underline{C}$ preserves kernels, then $\underline{C}$ is an abelian category with exact direct limits and a generator.

Proof. $\underline{C}$ is preadditive since it is a full subcategory of Mod-A . We have proved that $\underline{C}$ has limits and colimits. To prove that $\underline{C}$ is abelian, it only remains to show that if $\alpha: M \to N$ is a homomorphism in $\underline{C}$, then the canonical map

$$\bar{\alpha}: \mathrm{Coker}(\ker \alpha) \longrightarrow \mathrm{Ker}(\mathrm{coker}\ \alpha)$$

is an isomorphism. If we denote kernels and cokernels taken in Mod-A by underlining them, we have $\mathrm{Coker}(\ker \alpha) = a(\underline{\mathrm{Coker}}(\underline{\ker}\ \alpha))$ and $\mathrm{Ker}(\mathrm{coker}\ \alpha) = \underline{\mathrm{Ker}}(a(\underline{\mathrm{coker}}\ \alpha)) = a(\underline{\mathrm{Ker}}(\underline{\mathrm{coker}}\ \alpha))$ since a preserves kernels. $\bar{\alpha}$ is therefore an isomorphism.

Next we show that direct limits are exact. Let $\underline{D}$ be a small directed category and G , $G' : \underline{D} \to \underline{C}$ two functors with a monomorphism $G \to G'$. The induced morphism $\varinjlim iG \to \varinjlim iG'$ is a monomorphism in Mod-A , and since a preserves monomorphisms, it follows that $\varinjlim G \to \varinjlim G'$ is a monomorphism in $\underline{C}$. Finally, it is easy to see that a(A) will be a generator for $\underline{C}$.

Definition. A coreflective subcategory of Mod-A is called a Giraud subcategory if the left adjoint of the inclusion functor preserves kernels.

Thus if $i: \underline{C} \to$ Mod-A is a Giraud subcategory, then the left adjoint a of i is an exact functor. It is important to notice that the inclusion functor i is in general not exact; epimorphisms in $\underline{C}$ are not necessarily surjective maps. An abelian category with exact direct limits and a generator is usually called a Grothendieck category. Proposition 9.3 thus states that every Giraud subcategory is a Grothendieck category. Conversely, the Popescu-Gabriel theorem (Theorem 10.3) states that every Grothendieck category is a Giraud subcategory of Mod-A , where A is the endomorphism ring of some generator of the category.

References: Mitchell [102] (ch. V:5).

§ 10. Giraud subcategories and the Popescu-Gabriel theorem

Proposition 10.1. If $\underline{F}$ is a topology on A, then the $\underline{F}$-closed modules form a Giraud subcategory of Mod-A.

Proof. The category of $\underline{F}$-closed modules is coreflective by Proposition 7.10. Since the functors L and i preserve kernels and i is full and faithful, the relation $ia = LL$ implies that a preserves kernels.

Theorem 10.2. There is a 1-1 correspondence between topologies on A and equivalence classes of Giraud subcategories of Mod-A.

Proof. We already know how to associate a Giraud subcategory to a topology. Conversely, let $\underline{C}$ be a Giraud subcategory of Mod-A and let a be the left adjoint of the inclusion functor $i: \underline{C} \to$ Mod-A. Let $\underline{\underline{T}}$ be the class of modules M for which $a(M) = 0$. We verify that $\underline{\underline{T}}$ is a hereditary torsion class. Since a is exact, it is clear that $\underline{\underline{T}}$ is closed under extensions, submodules and quotient modules. Since a has a right adjoint, a commutes with direct sums and hence $\underline{\underline{T}}$ is closed also under direct sums. To $\underline{\underline{T}}$ there corresponds the topology $\{I \mid A/I \in \underline{\underline{T}}\}$, which is the topology we associate with the given Giraud subcategory $\underline{C}$.

We will now show that the two maps

$$\{\text{topologies on } A\} \underset{\Phi}{\overset{\Gamma}{\rightleftarrows}} \{\text{Giraud subcategories of Mod-A}\}$$

are the inverses of each other. We first show that $\Phi\Gamma = \text{id.}$. Let $\underline{F}$ be a toplogy and let $\underline{C}$ be the category of $\underline{F}$-closed modules. We must show that if $a(M) = 0$, then M is an $\underline{F}$-torsion module. But $M_{\underline{F}} = 0$ certainly implies $M = t(M)$.

It remains to show $\Gamma\Phi = \text{id.}$. Let $\underline{D}$ be a Giraud subcategory of Mod-A with the inclusion $i' : \underline{D} \to$ Mod-A and its left adjoint a' . Let $\underline{F}$ be the toplogy of right ideals I for which $a'(A/I) = 0$. We want to show that the category $\underline{C}$ of $\underline{F}$-closed modules is equivalent to $\underline{D}$.

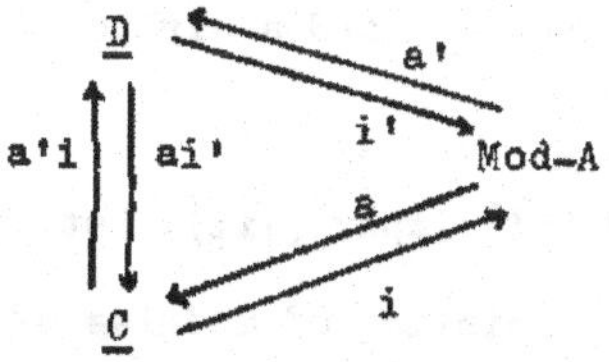

We first note that $a'ia \cong a'$, for if $M \in$ Mod-A then the exact sequence

$$0 \longrightarrow t(M) \longrightarrow M \longrightarrow M_{\underline{F}} \longrightarrow \text{Coker}\, \psi_M \longrightarrow 0$$

gives $a'(M) \cong a'(M_{\underline{F}})$. Similarly we have $ai'a' \cong a$, because the adjointness transformation $\phi : \text{id.} \to i'a'$ gives the exact sequence

$$0 \longrightarrow \text{Ker}\, \phi_M \longrightarrow M \xrightarrow{\phi_M} i'a'(M) \longrightarrow \text{Coker}\, \phi_M \longrightarrow 0$$

and $a'(\phi_M)$ is an isomorphism, so $\text{Ker}\, \phi_M$ and $\text{Coker}\, \phi_M$ are $\underline{F}$-torsion modules, and it follows that $a(\phi_M)$ is an isomorphism $a(M) \cong ai'a'(M)$.
From these two natural equivalences we obtain $a'i \cdot ai' \cong$ $\cong a'i' \cong \text{id.}$ and $ai' \cdot a'i \cong ai \cong \text{id.}$, and $\underline{C}$ and $\underline{D}$ are thus equivalent.

We may now state the Popescu-Gabriel theorem.

<u>Theorem 10.3.</u> Let $\underline{C}$ be a Grothendieck category with a generator U. Put $A = \mathrm{Hom}_{\underline{C}}(U,U)$ and let $T: \underline{C} \to \text{Mod-}A$ be the functor $T(C) = \mathrm{Hom}_{\underline{C}}(U,C)$. Then:

(i) T is full and faithful.

(ii) T has a left adjoint $S: \text{Mod-}A \to \underline{C}$ which is exact.

(iii) T induces an equivalence between $\underline{C}$ and a Giraud subcategory of Mod-A.

<u>Proof.</u> Let us first see how (i) and (ii) imply (iii). Let $\mathrm{Im}\, T$ be the full subcategory of Mod-A consisting of modules of the form $T(C)$, $C \in \underline{C}$. We have a commutative diagram

$$\begin{array}{ccc} \underline{C} & \xrightarrow{T} & \text{Mod-}A \\ & {\scriptstyle T'} \searrow \quad \nearrow {\scriptstyle i} & \\ & \mathrm{Im}\, T & \end{array}$$

and (i) states that T' is an equivalence. Define $a = T'S$: $\text{Mod-}A \to \mathrm{Im}\, T$. a is exact and is a left adjoint of i by (ii). $\mathrm{Im}\, T$ is thus a Giraud subcategory of Mod-A.

Proof of (i): $T = \mathrm{Hom}(U,\cdot)$ is faithful since U is a generator. To see that it is full, we must show that if $C, D \in \underline{C}$ and $\phi: \mathrm{Hom}_{\underline{C}}(U,C) \to \mathrm{Hom}_{\underline{C}}(U,D)$ is A-linear, then ϕ is of the form $\phi(f) = \varphi f$ for some $\varphi: C \to D$. Let $(f_i)_I$ be the set of all morphisms $U \to C$. There is a corresponding exact sequence

$$0 \longrightarrow K \xrightarrow{g} U^I \xrightarrow{f} C \longrightarrow 0$$

where U^I is the direct sum of I copies of U. The morphisms $\phi(f_i): U \to D$ induce a morphism $h: U^I \to D$. For each summand U_i of U^I we set $K_i = \mathrm{Ker}\, f_i = K \cap U_i$.

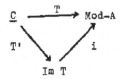

$$\begin{array}{ccccccccc}
0 & \longrightarrow & K_i & \xrightarrow{g_i} & U & \xrightarrow{f_i} & C & \longrightarrow & 0 \\
 & & \downarrow & & \downarrow{\scriptstyle u_i} & & \| & & \\
0 & \longrightarrow & K & \xrightarrow[g]{} & U^I & \xrightarrow[f]{} & C & \longrightarrow & 0 \\
 & & & & \downarrow{\scriptstyle h} & & & & \\
 & & & & D & & & &
\end{array}$$

For every $s \in \mathrm{Hom}(U,K_i)$ we get by the A-linearity of ϕ that $0 = \phi(f_i g_i s) = \phi(f_i) g_i s$, so $\phi(f_i) g_i = 0$ because U is a generator. It follows that $hg = 0$, and h factors as $h = \varphi f$ for some $\varphi : C \to D$. For each $f_i : U \to C$ we then have $\phi(f_i) = $ $= hu_i = \varphi f_i$.

Proof of the easy part of (ii), namely that a left adjoint S exists: For this we may either use general existence theorems for adjoints ([102], ch. V,§3), or we may proceed as follows. Consider A as a preadditive category $\hat{A}$ with only one object, and define in the obvious way a functor $u : \hat{A} \to \underline{C}$ with image object U. By a standard result in elementary category theory ([102], p.106) we can extend u to a colimit preserving functor $S : \text{Mod-}A \to \underline{C}$. Since every module M is the colimit of a functor $i \mapsto A_i$ $(i \in I)$, where each A_i is A_A, we get $\mathrm{Hom}_A(M,T(C)) = \mathrm{Hom}_A(\varinjlim A_i, \mathrm{Hom}_{\underline{C}}(U,C)) = \varprojlim \mathrm{Hom}_A(A_i, \mathrm{Hom}_{\underline{C}}(U,C)) = $ $= \varprojlim \mathrm{Hom}_{\underline{C}}(U_i,C) = \varprojlim \mathrm{Hom}_{\underline{C}}(S(A_i),C) = \mathrm{Hom}_{\underline{C}}(\varinjlim S(A_i),C) = $ $= \mathrm{Hom}_{\underline{C}}(S(\varinjlim A_i),C) = \mathrm{Hom}_{\underline{C}}(S(M),C)$, and thus S is a left adjoint of T.

Before we go on and prove the exactness of S, we make the following observation:

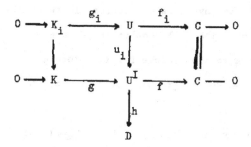

Proposition 10.4. Let $\underline{M}$ be any class of modules. There exists a strongest topology $\underline{F}$ such that all modules in $\underline{M}$ are $\underline{F}$-closed. A module L is a torsion module for this topology if and only if $\mathrm{Hom}(A,M) \xrightarrow{\cong} \mathrm{Hom}(\mathrm{Ann}(x),M)$ for every $x \in L$ and $M \in \underline{M}$.

Proof. If $\underline{E}$ is any topology, then a module M is $\underline{E}$-closed if and only if M is $\underline{E}$-torsion-free and $\underline{E}$-injective, which is equivalent to requiring M and $E(M)/M$ to be $\underline{E}$-torsion-free(Proposition 6.3). It follows that the torsion theory cogenerated by $\{ E(M) \oplus E(E(M)/M) \mid M \in \underline{M} \}$, which is hereditary by § 2, Exercise 4, defines the strongest topology for which all modules in $\underline{M}$ are closed.

It remains to determine the torsion modules for this strongest topology $\underline{F}$. If L is a torsion module, then $\mathrm{Ann}(x) \in \underline{F}$ for each $x \in L$ and hence $\mathrm{Hom}(A,M) \cong \mathrm{Hom}(\mathrm{Ann}(x),M)$ for all $M \in \underline{M}$. Conversely, if a module L satisfies this later condition, we may restate this as

$\mathrm{Hom}(C,M) = \mathrm{Ext}^1(C,M) = 0$ for every cyclic submodule C of L,

as one sees from the exact sequence

$$0 \to \mathrm{Hom}(A/\mathrm{Ann}(x),M) \to \mathrm{Hom}(A,M) \to \mathrm{Hom}(\mathrm{Ann}(x),M) \to \mathrm{Ext}^1(A/\mathrm{Ann}(x),M) \to 0$$

But if $\mathrm{Hom}(C,M) = 0$, then $\mathrm{Ext}^1(C,M) \cong \mathrm{Hom}(C,E(M)/M)$. Consequently we have $\mathrm{Hom}(C,M \oplus E(M)/M) = 0$ for every cyclic submodule C of L and $M \in \underline{M}$. This implies that L is a torsion module, because of the following easily verified fact:

Lemma 10.5. If L and M are modules, then $\mathrm{Hom}(L,E(M)) = 0$ if and only if $\mathrm{Hom}(C,M) = 0$ for every cyclic submodule C of L.

We continue the proof of Theorem 10.3, where it remains to show that S is exact. Let $\underline{F}$ be the strongest topology for which all modules T(C) , C ∈ $\underline{C}$, are $\underline{F}$-closed. Let $\underline{D}$ be the corresponding category of $\underline{F}$-closed modules. We have a diagram of functors

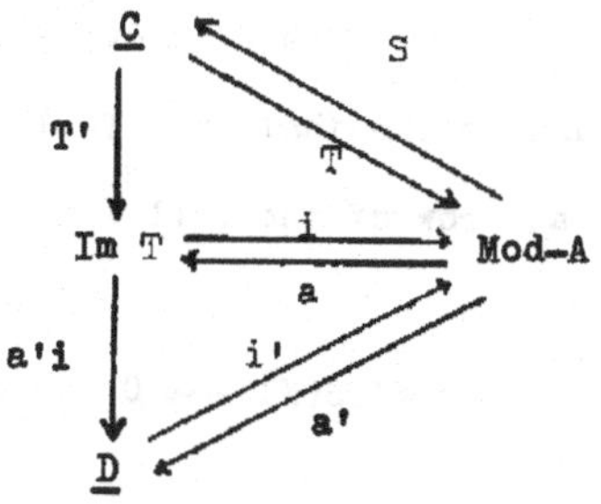

where T' is an equivalence. We have $i'a'i \cong i$ by the definition of $\underline{F}$. It now suffices to show that $a'iT' = a'T$ is an equivalence, because $T = i' \cdot a'T$ implies that the left adjoint S of T is the composition of the left adjoint a' of i' and the left adjoint of the equivalence a'T ; since both these two adjoints are exact, S will be exact. Since $i'a'T = T$ is full an faithful, also a'T is full and faithful, and it remains for us to show that every module in $\underline{D}$ is isomorphic to a module of the form a'T(C) , i.e. that every $\underline{F}$-closed module M is isomorphic to T(C).

Thus let M be an $\underline{F}$-closed module. Choose an exact sequence

$$(*) \qquad \bigoplus_I A \xrightarrow{(\alpha_{ij})} \bigoplus_J A \longrightarrow M \longrightarrow 0$$

in Mod-A . Since the functor S has a right adjoint, it preserves colimits and therefore carries (*) into an exact sequence

$$(**)\quad \bigoplus_I U \xrightarrow{S(\alpha_{ij})} \bigoplus_J U \longrightarrow S(M) \longrightarrow 0$$

in $\underline{C}$.

Lemma 10.6. The functor $a'T:\underline{C} \to \underline{D}$ is exact and preserves direct sums.

We conclude the proof of the theorem before we prove the lemma. By applying the Lemma to $(**)$ and noting that $A = T(U)$ is $\underline{F}$-closed, we obtain the upper exact row of the following diagram in $\underline{D}$:

$$\begin{array}{ccccccc} \bigoplus_I A & \xrightarrow{a'TS(\alpha_{ij})} & \bigoplus_J A & \longrightarrow & a'TS(M) & \longrightarrow & 0 \\ \| & & \| & & & & \\ \bigoplus_I A & \xrightarrow[a'(\alpha_{ij})]{} & \bigoplus_J A & \longrightarrow & M & \longrightarrow & 0 \end{array}$$

The lower row is obtained by applying a' to $(*)$, and is also exact. The diagram commutes because $A = T(U)$ implies that α_{ij} has the form $\alpha_{ij} = i(\beta_{ij})$, and one has $a'TSi = a'iT'Si =$ $= a'iai = a'i$. We conclude from the diagram that $M \cong a'TS(M)$.

Proof of Lemma 10.6: We already know the functor $a'T$ to be left exact, so to prove exactness it will suffice to show that it preserves epimorphisms. This means that if $f:C' \to C''$ is an epimorphism in $\underline{C}$, then we should show that $\operatorname{Coker} T(f)$ is an $\underline{F}$-torsion module. By Proposition 10.4 this is equivalent to showing that for each $x \in T(C'')$ we have

$$\operatorname{Hom}(A,T(C)) \cong \operatorname{Hom}((\operatorname{Im} T(f) :x),T(C))$$

for all $C \in \underline{C}$. Define $h:U \to C''$ such that $T(h):A \to T(C'')$ maps 1 into x . From the pullback in $\underline{C}$

$$\begin{array}{ccccccccc} 0 & \rightarrow & K & \xrightarrow{k} & P & \xrightarrow{g} & U & \rightarrow & 0 \\ & & \| & & \downarrow & & \downarrow h & & \\ 0 & \rightarrow & K & \longrightarrow & C' & \xrightarrow[f]{} & C'' & \rightarrow & 0 \end{array}$$

we get a pullback diagram with exact rows in Mod-A

$$\begin{array}{ccccccc} 0 & \rightarrow & T(K) & \xrightarrow{T(k)} & T(P) & \xrightarrow{T(g)} & A \\ & & \| & & \downarrow & & \downarrow \\ 0 & \rightarrow & T(K) & \longrightarrow & T(C') & \xrightarrow[T(f)]{} & T(C'') \end{array}$$

and $(\operatorname{Im} T(f) : x) = \operatorname{Im} T(g)$. Note that A is the cokernel of $T(k)$ in the subcategory $\operatorname{Im} T$, because $T' : \underline{C} \rightarrow \operatorname{Im} T$ is an equivalence. It follows that if we have a homomorphism $\operatorname{Im} T(g) \rightarrow T(C)$, for some $C \in \underline{C}$, then it factors uniquely over $A = \operatorname{Coker} T(k)$, and this is precisely what we wanted to show.

It remains to show that $a'T$ preserves direct sums. Actually we prove a little more, namely that $a'T$ preserves directed unions. Let (C_α) be a directed family of subobjects of $C \in \underline{C}$. We must show that the cokernel of the monomorphism

$$f : \bigcup_\alpha T(C_\alpha) \longrightarrow T(\bigcup_\alpha C_\alpha)$$

is a torsion module. By Proposition 10.4 this means that for each $x \in T(\bigcup C_\alpha)$ we shall show that $\operatorname{Hom}(A, T(C')) \cong \operatorname{Hom}((\operatorname{Im} f : x), T(C'))$ for all $C' \in \underline{C}$. Define $h : U \rightarrow \bigcup C_\alpha$ such that $T(h) : A \rightarrow T(\bigcup C_\alpha)$ maps 1 to x . From the pullback diagram

$$\begin{array}{ccc} P_\alpha & \hookrightarrow & U \\ \downarrow & & \downarrow h \\ C_\alpha & \hookrightarrow & \bigcup C_\alpha \end{array}$$

we obtain a commutative diagram

$$\begin{array}{ccc} \bigcup T(P_\alpha) & \xrightarrow{g} & A \\ \downarrow & & \downarrow T(h) \\ \bigcup T(C_\alpha) & \xrightarrow[f]{} & T(\bigcup C_\alpha) \end{array}$$

which is a pullback diagram because pullbacks are preserved both by T and when taking direct limits in Mod-A. Therefore we have $(\mathrm{Im}\ f : x) = \mathrm{Im}\ g \cong \bigcup T(P_\alpha)$. Now $\mathrm{Hom}(\bigcup T(P_\alpha), T(C')) \cong$ $\cong \varprojlim \mathrm{Hom}(T(P_\alpha), T(C')) \cong \varprojlim \mathrm{Hom}(P_\alpha, C') \cong \mathrm{Hom}(\bigcup P_\alpha, C') =$ $= \mathrm{Hom}(U, C') \cong \mathrm{Hom}(A, T(C'))$, where we have utilized the fact that exactness of $\varinjlim$ implies $\bigcup P_\alpha = U$.

<u>Example</u>: Let A be any ring. Proposition 10.4 provides A with a strongest topology $\underline{F}$ for which A is $\underline{F}$-closed. This topology is called the <u>canonical topology</u> on A.

<u>Exercises</u>:

1. Let $\underline{F}$ be a topology on Mod-A and let T:Mod-$A \to \underline{B}$ be an exact functor into an abelian category $\underline{B}$, such that $T(M) = 0$ for all $\underline{F}$-torsion modules. Show that T has a unique factorization $T = T'a$ over the category of $\underline{F}$-closed modules. (The category of $\underline{F}$-closed modules is thus a solution of a universal problem).
2. Let $\underline{C}$ be a Giraud subcategory of Mod-A and let a be the left adjoint of the inclusion functor i. Show that:
   (i) If E is an injective object in $\underline{C}$, then $i(E)$ is an injective module.
   (ii) i preserves injective envelopes.

(iii) If the torsion theory corresponding to $\underline{C}$ is stable (§ 4) and E is an injective module, then a(E) is injective in $\underline{C}$.

<u>References</u>: Bucur-Deleanu [12] (ch. 6 , written by N. Popescu), Gabriel [31] (ch. 3), Lambek [48] (§ 4), Popescu-Gabriel [62], Roos [66] (ch. 1), Takeuchi [112].

Chapter 3. General properties of rings of quotients

§ 11. Lattices of $\underline{F}$-pure submodules

We will assume $\underline{F}$ to be a topology on A. For each module M and submodule L of M, we define

$$L^c = \{ x \in M \mid (L:x) \in \underline{F} \} .$$

The operation $L \mapsto L^c$ is a closure operation on the lattice of all submodules of M. Those submodules L for which $L^c = M$ were called $\underline{F}$-submodules of M in § 6. On the other hand, we put

$$\underline{C}_{\underline{F}}(M) = \{ L \subset M \mid L^c = L \} .$$

Note that $L^c = L$ if and only if M/L is torsion-free; in particular, if M is torsion-free, then $\underline{C}_{\underline{F}}(M)$ is the family of $\underline{F}$-pure submodules of M (§ 6, Exercise).

Proposition 11.1. $\underline{C}_{\underline{F}}(M)$ is a complete modular lattice.

Proof. From the fact that $L \mapsto L^c$ is a closure operation it follows that $\underline{C}_{\underline{F}}(M)$ is a complete lattice with intersection as meet ([18], Ch. 2.1). The join is given by $\bigvee L_i = (\Sigma L_i)^c$. It remains to verify modularity. Let H, K and L be members of $\underline{C}_{\underline{F}}(M)$ with $H \subset K$. Then $K \cap (H \vee L) = K^c \cap (H + L)^c = (K \cap (H + L))^c = (H + (K \cap L))^c = H \vee (K \cap L)$, using the modularity of the lattice of all submodules of M.

Proposition 11.2. If $K \in \underline{F}(M)$, then there is a lattice isomorphism $\underline{C}_{\underline{F}}(M) \to \underline{C}_{\underline{F}}(K)$ given by $L \mapsto L \cap K$.

Proof. If $L \in \underline{C}_{\underline{F}}(M)$, then clearly $L \cap K \in \underline{C}_{\underline{F}}(K)$. The inverse map is defined as $L \mapsto L^c$, where the closure is taken in M.

For if $L \in \underline{C}_{\underline{F}}(K)$, then $L^c \cap K = \{x \in K \mid (L:x) \in \underline{F}\} = L$, while if $L \in \underline{C}_{\underline{F}}(M)$, then $(L \cap K)^c = L^c \cap K^c = L \cap M = L$.

In the following we will mainly be concerned with $\underline{C}_{\underline{F}}(M)$ when M is torsion-free. In case M is also $\underline{F}$-injective, we have:

<u>Proposition 11.3.</u> Let M be $\underline{F}$-closed. A submodule L of M is $\underline{F}$-closed if and only if $L \in \underline{C}_{\underline{F}}(M)$.

<u>Proof.</u> If $L \subset M$, then M/L is torsion-free if and only if L is $\underline{F}$-closed, as is seen from the exact sequence

$$0 = \mathrm{Hom}(A/I,M) \to \mathrm{Hom}(A/I,M/L) \to \mathrm{Ext}^1(A/I,L) \to \mathrm{Ext}^1(A/I,M) = 0 .$$

Recall that a submodule L of M is called <u>complemented</u> if L is maximal with respect to $L \cap K = 0$ for some $K \subset M$ (§ 4).

<u>Proposition 11.4.</u> Every complemented submodule of a torsion-free module M is a member of $\underline{C}_{\underline{F}}(M)$.

<u>Proof.</u> If L is maximal such that $L \cap K = 0$, then $L^c \cap K^c = (L \cap K)^c = 0^c = 0$ since M is torsion-free. We must then have $L^c = L$ by maximality.

<u>Proposition 11.5.</u> The following properties of a torsion-free module M are equivalent:

(a) $\underline{C}_{\underline{F}}(M)$ is a complemented lattice.

(b) $\underline{C}_{\underline{F}}(M)$ consists of the complemented submodules of M .

(c) Every essential submodule of M is an $\underline{F}$-submodule.

<u>Proof.</u> (a) $\Rightarrow$ (b): Every complemented submodule is in $\underline{C}_{\underline{F}}(M)$ by 11.4. Suppose conversely $L \in \underline{C}_{\underline{F}}(M)$. By hypothesis there exists

$K \subset M$ such that $K \cap L = 0$ and $(K + L)^c = M$. Let $L' \supset L$ be maximal with respect to $L' \cap K = 0$. For each $x \in L'$ we have $xI \subset K + L$ for some $I \in \underline{F}$. But $xA \cap K = 0$, so $xI \subset K$. $L \in \underline{C}_{\underline{F}}(M)$ implies $x \in L$ and hence $L = L'$ is complemented.

(b) $\Rightarrow$ (c): Let L be an essential submodule of M. L^c is then both complemented and essential in M. Hence $L^c = M$, and L is an $\underline{F}$-submodule.

(c) $\Rightarrow$ (a): If $L \in \underline{C}_{\underline{F}}(M)$, choose K maximal with respect to $K \cap L = 0$. $K + L$ is then an essential submodule of M, hence an $\underline{F}$-submodule. Thus we have $K \vee L = M$ and $K \cap L = 0$.

Proposition 11.6. Let M be an $\underline{F}$-closed module such that $\underline{C}_{\underline{F}}(M)$ is complemented. Then:

(i) Every $\underline{F}$-closed submodule of M is a direct summand.

(ii) The endomorphism ring of M is regular (in the sense of von Neumann).

Proof. (i): If $L \subset M$ is $\underline{F}$-closed, then $L \in \underline{C}_{\underline{F}}(M)$ by 11.3. Hence there exists $K \subset M$ such that $K \cap L = 0$ and $K + L$ is essential, hence an $\underline{F}$-submodule, in M. By Proposition 6.2 we may extend the canonical projection $K + L \rightarrow L$ to a homomorphism $M \rightarrow L$. This makes L into a direct summand of M.

(ii): Let $f: M \rightarrow M$ be an endomorphism. Then $\mathrm{Ker}\, f \in \underline{C}_{\underline{F}}(M)$, because if $x \in M$ and $xI \subset \mathrm{Ker}\, f$ for some $I \in \underline{F}$, then $f(x)I = 0$ and M torsion-free implies $x \in \mathrm{Ker}\, f$. $\mathrm{Ker}\, f$ is thus a direct summand of M. Write $M = \mathrm{Ker}\, f \oplus K$. f induces an isomorphism $f|K: K \rightarrow \mathrm{Im}\, f$, so also $\mathrm{Im}\, f$ is an $\underline{F}$-closed

module. Hence Im f is a direct summand of M , and we may
therefore extend $(f|K)^{-1}$ to a homomorphism h:M→ M . Then
f = fhf , and we have established the regularity of the
endomorphism ring.

The lattice $\underline{C}_{\underline{F}}(A)$ has an interesting description as the
set of annihilators of subsets of an injective module. Recall
that $\underline{C}_{\underline{F}}(A) = \{ I \mid A/I$ is torsion-free $\}$. In the following
three propositions we let E be an injective module which
cogenerates the torsion theory corresponding to $\underline{F}$, i.e.
$\underline{F} = \{ I \mid \text{Hom}(A/I,E) = 0 \}$. If S is a subset of any module,
we put $\text{Ann}(S) = \{ a \in A \mid Sa = 0 \}$.

Proposition 11.7. $\underline{C}_{\underline{F}}(A) = \{ \text{Ann}(S) \mid$ subsets $S \subseteq E \}$.
Proof. Suppose $S \subseteq E$. Then $\text{Ann}(S) = \bigcap_{x \in S} \text{Ann}(x)$, where each
$\text{Ann}(x) \in \underline{C}_{\underline{F}}(A)$ since $A/\text{Ann}(x) \subseteq E$ is torsion-free. But $\underline{C}_{\underline{F}}(A)$
is closed under intersections, so also $\text{Ann}(S) \in \underline{C}_{\underline{F}}(A)$.

Suppose on the other hand that $I \in \underline{C}_{\underline{F}}(A)$ and put
$S = \{ x \in E \mid xI = 0 \}$. Then $\text{Ann}(S) \supseteq I$. To show that every
$a \in \text{Ann}(S)$ belongs to I , it suffices to show that $(I:a) \in \underline{F}$,
i.e. that $\text{Hom}(A/(I:a),E) = 0$. Let $\alpha:A/(I:a) \to A/I$ be the
monomorphism $\alpha(\bar{b}) = \overline{ab}$. For every $f:A/(I:a) \to E$ we get a
commutative diagram

$$
\begin{array}{ccc}
A/(I:a) & \overset{\alpha}{\hookrightarrow} & A/I \\
f \downarrow & \swarrow g & \\
E & &
\end{array}
$$

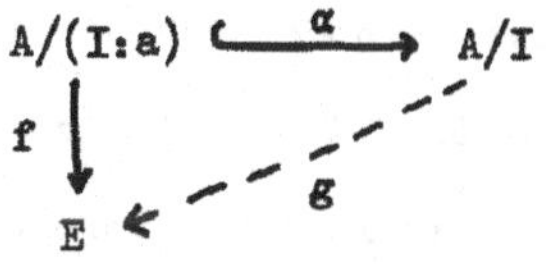

where $g(\bar{b}) = xb$ for some $x \in E$. Then necessarily $xI = 0$, so $x \in S$. But then also $xa = 0$, so $g\alpha = 0$. Hence $f = 0$.

Before the statement of the next result we need to make two definitions. A lattice is called noetherian if every ascending chain is stationary. An injective module is called Σ-injective if every direct sum of copies of the module is injective.

Proposition 11.8. The following assertions are equivalent:

(a) Every direct sum of torsion-free injective modules is injective.

(b) E is Σ-injective.

(c) $\underline{C}_{\underline{F}}(A)$ is a noetherian lattice.

Proof. (a) $\Rightarrow$ (b) is trivial.

(b) $\Rightarrow$ (c): By 11.7 it suffices to show that every strictly ascending chain $I_1 \subset I_2 \subset \ldots$ of annihilators of subsets of E must be finite. If the chain were not finite, we could choose for each n an element $x_n \in E$ such that $x_n I_n = 0$ but $x_n I_{n+1} \neq 0$. Put $I = \bigcup_1^\infty I_n$ and define $f: I \to \bigoplus_1^\infty E$ as $f(a) = (x_1 a, x_2 a, \ldots)$. Note that f is well-defined! Since E is Σ-injective, f has the form $f(a) = ya$ for some $y = (y_1, y_2, \ldots, y_m, 0, \ldots) \in \bigoplus_1^\infty E$, which contradicts the choice of the elements x_n .

(c) $\Rightarrow$ (a): Let (E_α) be a family of torsion-free injective modules. Suppose we are given a homomorphism $f: I \to \bigoplus_\alpha E_\alpha$. It suffices to show that f maps I into the sum of finitely many E_α . Suppose on the contrary that there exists an infinite sequence of indices α , which we write as $\alpha = 1, 2, \ldots$, such that $\operatorname{Im} f$ has non-zero coordinates in each E_n . Put $I_n = f^{-1}(E_1 \oplus \dots \oplus E_n)$. The ascending chain $I_1^c \subset I_2^c \subset \dots$ is by hypothesis finite, which implies that for some n we have $I_k \subset I_n^c$ for all k . Let $a \in I_k$ be some element for which $f(a)$ has non-zero coordinate x_k in E_k , for some $k > n$. Since $a \in I_n^c$, we have $aJ \subset I_n$ for some $J \in \underline{F}$. This gives $f(a)J \subset E_1 \oplus \dots \oplus E_n$. But then $x_k J = 0$, which contradicts the assumption that E_k is torsion-free.

<u>Proposition 11.9.</u> If $\underline{C}_{\underline{F}}(A)$ is noetherian, then $\underline{F}$ contains a cofinal family of finitely generated right ideals.

<u>Proof.</u> For each right ideal I of A we set $I^{\perp} = \{ x \in E \mid xI = 0 \}$. Note that $I \in \underline{F}$ if and only if $I^{\perp} = 0$. The operations Ann and $^{\perp}$ define an order-inverting bijection between the set of right ideals of the form $\operatorname{Ann}(S)$ and the set of submodules of E of the form $I^{\perp}$. Since we have ACC on the former set (by 11.7), we must have DCC on the submodules $I^{\perp}$. Let I be any right ideal in $\underline{F}$. Consider the family $\{ J^{\perp} \mid J$ finitely generated right ideal $\subset I \}$ and let J be a minimal member of this family. For each $a \in I$, the right ideal $J_1 = J + aA$ is also finitely generated $\subset I$ and satisfies $J_1^{\perp} \subset J^{\perp}$. By minimality we must have $J_1^{\perp} = J^{\perp}$ so in particular $J^{\perp}a = 0$. Since this holds for all $a \in I$, we have $J^{\perp}I = 0$ and thus $J^{\perp} \subset I^{\perp}$, which implies that also $J^{\perp} = 0$ and $J \in \underline{F}$.

We will now show that $\underline{C}_{\underline{F}}(A)$ is isomorphic to a corresponding lattice of right ideals in $A_{\underline{F}}$, assuming for simplicity that A is torsion-free, so that A is a subring of $A_{\underline{F}}$. Define

$$\underline{F}^e = \{\text{right ideals } J \text{ of } A_{\underline{F}} \mid J \cap A \in \underline{F}\}.$$

<u>Proposition 11.10.</u> $\underline{F}^e$ is a topology on $A_{\underline{F}}$.

<u>Proof.</u> Suppose $J \in \underline{F}^e$ and $q \in A_{\underline{F}}$. Then $(J:q) \cap A = \{a \in A \mid qa \in J\} \supset \{a \in A \mid qa \in J \cap A\} \in \underline{F}$, because $J \cap A \in \underline{F}$ and A is an $\underline{F}$-submodule of $A_{\underline{F}}$, so we may use Lemma 6.1. Thus T 1 is satisfied. Next we verify T 2. Let I be a right ideal of $A_{\underline{F}}$ such that there exists $J \in \underline{F}^e$ for which $(I:q) \in \underline{F}^e$ for all $q \in J$. Then $J \cap A \in \underline{F}$ and for each $a \in J \cap A$ we have $\{b \in A \mid ab \in I \cap A\} = \{b \in A_{\underline{F}} \mid ab \in I\} \cap A \in \underline{F}$. T 2 for $\underline{F}$ implies $I \cap A \in \underline{F}$ and hence $I \in \underline{F}^e$.

<u>Proposition 11.11.</u> $A_{\underline{F}}$ is equal to its ring of quotients with respect to $\underline{F}^e$.

<u>Proof.</u> The ring of quotients of $A_{\underline{F}}$ is isomorphic to the $\underline{F}^e$-injective envelope of $A_{\underline{F}}$, and is a submodule of $E(A_{\underline{F}}) = E(A)$ (Propositions 8.1 and 8.2). But the $\underline{F}^e$-injective envelope of $A_{\underline{F}}$ coincides with its $\underline{F}$-injective envelope, as one immediately verifies, and $A_{\underline{F}}$ is $\underline{F}$-closed.

<u>Proposition 11.12.</u> If M is a torsion-free A-module, then $\underline{C}_{\underline{F}}(M) \cong \underline{C}_{\underline{F}^e}(M_{\underline{F}})$.

<u>Proof.</u> Since M is an $\underline{F}$-submodule of $M_{\underline{F}}$, we have $\underline{C}_{\underline{F}}(M) \cong \underline{C}_{\underline{F}}(M_{\underline{F}})$ by 11.2. It remains to see that $\underline{C}_{\underline{F}}(M_{\underline{F}}) = \underline{C}_{\underline{F}^e}(M_{\underline{F}})$.

Suppose $L \in \underline{C}_{\underline{F}}(M_{\underline{F}})$. L is then an $\underline{F}$-closed module by 11.3 and is therefore an $A_{\underline{F}}$-module. If $x \in M_{\underline{F}}$ has the property that $xJ \subset L$ for some $J \in \underline{F}^e$, then $x(J \cap A) \subset L$ and hence $x \in L$. Consequently $L \in \underline{C}_{\underline{F}^e}(M_{\underline{F}})$.

Suppose conversely that $L \in \underline{C}_{\underline{F}^e}(M_{\underline{F}})$. If $x \in M_{\underline{F}}$ and $xI \subset L$ for some $I \in \underline{F}$, then $xIA_{\underline{F}} \subset L$ and hence $IA_{\underline{F}} \in \underline{F}^e$ implies $x \in L$. It follows that $L \in \underline{C}_{\underline{F}}(M_{\underline{F}})$.

Examples:

1. Let $\underline{F}$ be the Goldie topology, i.e the topology generated by the family of essential right ideals (§ 3, Example 1). From Propositions 11.5 and 11.6 we obtain:

Proposition 11.13. The endomorphism ring of a non-singular injective module is regular.

For the Goldie topology one can prove the converse of Proposition 11.9, namely:

Proposition 11.14. The following assertions are equivalent for the Goldie topology $\underline{F}$:

(a) Every direct sum of non-singular injective modules is injective.

(b) The lattice $\underline{C}_{\underline{F}}(A)$ of complemented right ideals is noetherian.

(c) $\underline{F}$ contains a cofinal family of finitely generated right ideals.

Proof. It remains to prove (c) $\Rightarrow$ (a). Let (E_α) be a family of non-singular injective modules, and let $f: I \to \bigoplus_\alpha E_\alpha$ be a homomorphism, for an arbitrary right ideal I. Choose a right ideal J such that $I \cap J = 0$ and $I + J$ is essential in A. By (c) there exists a finitely generated right ideal $K \in \underline{F}$ contained in $I + J$. Extend f to $I \oplus J \to \bigoplus_\alpha E_\alpha$ by $f|J = 0$ and then restrict to a homomorphism $g: K \to \bigoplus_\alpha E_\alpha$. Since K is finitely generated, g maps into the sum of finitely many E_α, and is therefore extendable to $h: A \to \bigoplus_\alpha E_\alpha$. Since $\bigoplus_\alpha E_\alpha$ is torsion-free, the usual argument shows that $h|I = f$ (cf. the proof of Lemma 8.3).

2. Taking $\underline{F} = \{A\}$, we otain as a special case of Proposition 11.8 that A is right noetherian if and only if every direct sum of injective modules is injective.

3. Let $\underline{D}$ be the family of dense right ideals of A (§ 3, Example 2). Then $\underline{D}^e$ is the family of dense right ideals of Q_m, as one easily verifies by means of Proposition 3.8.

Exercises:

1. Let A be a regular ring and let $\underline{F}$ be the family of essential right ideals. Show that:
   (i) A is non-singular.
   (ii) $\underline{C}_{\underline{F}}(A)$ is noetherian if and only if A is semi-simple.
2. Show that the following two properties of a topology $\underline{F}$ are equivalent:
   (a) If $I_1 \subset I_2 \subset \dots$ is a countable ascending chain of

right ideals such that $\bigcup_1^\infty I_n \in \underline{F}$, then some $I_n \in \underline{F}$.

(b) If $I_1 \subset I_2 \subset \ldots$ is a countable ascending chain in $\underline{C}_{\underline{F}}(A)$, then also $\bigcup_1^\infty I_n \in \underline{C}_{\underline{F}}(A)$.

Show that these properties are satisfied if every $I \in \underline{F}$ contains a finitely generated $J \in \underline{F}$.

3. Show that the following properties of a right self-injective ring A are equivalent:

   (a) A satisfies ACC on right annihilators of subsets of A .

   (b) A_A is Σ-injective.

   (c) Every projective module is injective.

4. The ring A is called right finite-dimensional if no right ideal can be written as a direct sum of infinitely many non-zero right ideals of A . Show that:

   (i) A is right finite-dimensional if and only if every right ideal is an essential extension of a finitely generated right ideal.

   (ii) Every right finite-dimensional ring satisfies the conditions of Proposition 11.14. (Hint: use 3.6 to verify 11.14(c)).

5. Show that if E is an injective module and M is non-singular, then every exact sequence $0 \to K \to E \to M \to 0$ splits. Using this, show that if E and E' are injective submodules of a non-singular module, then also E + E' is injective.

6. Show that the conditions of Proposition 11.14 are equivalent to:

   (d) Every non-singular module contains a unique maximal injective submodule.

7. Let $\underline{F}$ be a topology on A. Show that $\underline{F}^e$ is the strongest topology $\underline{F}'$ on $A_{\underline{F}}$ such that all $\underline{F}$-closed modules (considered as $A_{\underline{F}}$-modules) are $\underline{F}'$-closed.

References: Armendariz [85], Faith [27], [28] (§7 and 8), Johnson [40], Teply [75],[76], Utumi [80].

§ 12. Finiteness conditions on topologies

In this § we will consider two kinds of finiteness conditions on the topology $\underline{F}$. The first one is introduced in the next proposition, where $\underline{C}$ as usual denotes the category of $\underline{F}$-closed modules and $i:\underline{C} \to \text{Mod-}A$ is the inclusion functor, while $q:\text{Mod-}A \to \text{Mod-}A_{\underline{F}}$ is the functor $M \mapsto M_{\underline{F}}$.

Proposition 12.1. The following assertions are equivalent:

(a) If $I_1 \subset I_2 \subset \ldots$ is a countable ascending chain of right ideals such that $\bigcup_1^{\infty} I_k \in \underline{F}$, then $I_n \in \underline{F}$ for some n.

(b) Every direct sum of $\underline{F}$-closed modules is $\underline{F}$-closed.

(b') Every direct sum of countably many $\underline{F}$-closed modules is $\underline{F}$-closed.

(c) The functor i commutes with direct sums.

(d) The functor q commutes with direct sums.

Proof. (a) $\Rightarrow$ (b): Let $\{M_\alpha\}$ be a family of $\underline{F}$-closed modules.

$\oplus M_\alpha$ is of course torsion-free, and we must show that it also is $\underline{F}$-injective. Let $f: I \to \oplus M_\alpha$ be any homomorphism with $I \in \underline{F}$. Considering $\oplus M_\alpha$ as a submodule of the $\underline{F}$-injective module $\prod M_\alpha$, there exists $x = (x_\alpha) \in \prod M_\alpha$ such that $f(a) = xa$ for all $a \in I$. We only have to show that $x \in \oplus M_\alpha$. If this were not true, there would exist an infinite set $\{\alpha_1, \alpha_2, \ldots\}$ of indices α for which $x_{\alpha_i} \neq 0$. Put $I_n = \{ a \in I \mid x_{\alpha_i} a = 0$ for $i \geq n \}$. Since $f(I) \subset \oplus M_\alpha$, we have $I = \bigcup\limits_n I_n$. But (a) then implies that $I = I_n$ for some n. Now M_{α_n} is torsion-free, so $x_{\alpha_n} I = 0$ implies $x_{\alpha_n} = 0$, which is a contradiction.

(b) $\Rightarrow$ (c): Let (M_α) be a family of $\underline{F}$-closed modules. The direct sum of the modules M_α in the category $\underline{C}$ of $\underline{F}$-closed modules is $\oplus M_\alpha = a(\oplus i(M_\alpha))$ by Proposition 9.2. So if $\oplus i(M_\alpha)$ is $\underline{F}$-closed, then $i(\oplus M_\alpha) = \oplus i(M_\alpha)$.

(c) $\Rightarrow$ (d) $\Rightarrow$ (b) and (b) $\Rightarrow$ (b') are rather obvious.

(b') $\Rightarrow$ (a): Let $I_1 \subset I_2 \subset \ldots$ be an ascending chain with $I = \bigcup I_n \in \underline{F}$. There is a well-defined canonical map $I \to \oplus A/I_n$. Since $I \in \underline{F}$, one obtains a commutative diagram

$$\begin{array}{ccc} I & \longrightarrow & A \\ \downarrow & & \downarrow f \\ \oplus A/I_n & \longrightarrow & \oplus (A/I_n)_{\underline{F}} \end{array}$$

where $f(a) = xa$ for some $x = (x_n) \in \oplus (A/I_n)_{\underline{F}}$. There exists m such that $x_n = 0$ for $n \geq m$. The image of I in A/I_m then lies in the kernel of $A/I_m \to (A/I_m)_{\underline{F}}$, so I/I_m is a torsion module. The exact sequence

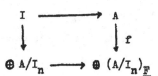

$$0 \longrightarrow I/I_m \longrightarrow A/I_m \longrightarrow A/I \longrightarrow 0$$

where also A/I is torsion, shows that A/I_m is torsion, and hence $I_m \in \underline{F}$.

We strengthen the finiteness condition somewhat by considering topologies with the following properties:

Proposition 12.2. The following assertions are equivalent for a topology $\underline{F}$:

(a) $\underline{F}$ contains a cofinal family of finitely generated right ideals.

(b) Every directed union of $\underline{F}$-closed modules is $\underline{F}$-closed.

(c) The functor i commutes with directed unions.

(d) The torsion radical t commutes with direct limits.

Proof. (a) $\Rightarrow$ (b): Let (M_α) be a directed family of $\underline{F}$-closed submodules of some module. $\bigcup_\alpha M_\alpha$ is then torsion-free, so it remains to verify that it is $\underline{F}$-injective. Consider any homomorphism $f: I \to \bigcup M_\alpha$ where $I \in \underline{F}$. I contains a finitely generated $J \in \underline{F}$. f maps J into some M_α, so by the $\underline{F}$-injectivity of M_α there exists $x \in M_\alpha$ such that $f(a) = xa$ for $a \in J$. Since $\bigcup M_\alpha$ is torsion-free, one then has $f(a) = xa$ also for all $a \in I$ (by the same argument as in the proof of Proposition 7.8).

(b) $\Leftrightarrow$ (c) similarly to the preceding Proposition.

(b) $\Rightarrow$ (a): Write $I \in \underline{F}$ as the directed union of finitely generated right ideals I_α. Then $ia(A) = ia(I) = ia(\bigcup I_\alpha) =$

$= \bigcup ia(I_\alpha)$, and so the canonical homomorphism $\psi: A \to A_{\underline{F}}$ factors as

$$\begin{array}{ccc} A & \xrightarrow{\psi} & ia(A) \\ {\scriptstyle f}\downarrow & & \downarrow{\scriptstyle \sigma}\ \cong \\ ia(I_\alpha) & \overset{j}{\hookrightarrow} & \bigcup ia(I_\alpha) \end{array}$$

for some α . Let $g: ia(I_\alpha) \to ia(A)$ be the canonical map. We want to show that g is an isomorphism, because this would imply $I_\alpha \in \underline{F}$. g is obviously a monomorphism, since $\sigma g = j$. We have $\sigma gf = jf = \sigma\psi$, so $gf = \psi$. It follows that $\operatorname{Im} g$ is an $\underline{F}$-closed submodule of $A_{\underline{F}}$ containing $\operatorname{Im}\psi$, and we conclude that g is an epimorphism.

(a) $\Rightarrow$ (d): Let (M_α) be a direct system of modules. The inclusions $t(M_\alpha) \to M_\alpha$ induce in the limit an inclusion $\varinjlim t(M_\alpha) \to \varinjlim M_\alpha$. The class of torsion modules is closed under direct limits, since it is closed under direct sums and quotients. $\varinjlim t(M_\alpha)$ is therefore a submodule of $t(\varinjlim M_\alpha)$. To show that we actually have equality, suppose $x \in t(\varinjlim M_\alpha)$. Then $xI = 0$ for some finitely generated $I \in \underline{F}$. Since I is finitely generated, it is clear that we may represent x by some $x_\alpha \in M_\alpha$ such that still $x_\alpha I = 0$. Then $x_\alpha \in t(M_\alpha)$, and $x \in \varinjlim t(M_\alpha)$.

(d) $\Rightarrow$ (a): Write $I \in \underline{F}$ as the directed union of finitely generated right ideals I_α . $A/I = \varinjlim A/I_\alpha$ is a torsion module, so $A/I = t(A/I) = \varinjlim t(A/I_\alpha)$. In particular the generator $\bar{1} \in A/I$ comes from some $t(A/I_\alpha)$, which means that there exist $a \in A$ and $J \in \underline{F}$ such that $aJ \subset I_\alpha$ and $1-a \in I$. We may choose α so that $1-a \in I_\alpha$, and then $J \subset I_\alpha$. Hence $I_\alpha \in \underline{F}$.

It is clear that every topology satisfying Proposition 12.2 also satisfies 12.1 (cf. Exercise 2 of § 11). The converse holds e.g. when all right ideals in A are countably generated.

References: Goldman [33], Roos [66] (ch. 1).

§ 13. Flat epimorphisms of rings

In many examples of rings of quotients one obtains the module $M_{\underline{F}}$ of quotients as $M_{\underline{F}} = M \otimes_A A_{\underline{F}}$. Here we will prove that this is equivalent to several other nice properties of the localization, e.g. that $A_{\underline{F}}$ is obtained by a kind of generalized calculus of fractions.

Let $\underline{F}$ be a topology on A and let $\psi : A \to A_{\underline{F}}$ be the canonical ring homomorphism. We have the diagram of functors (cf. § 7):

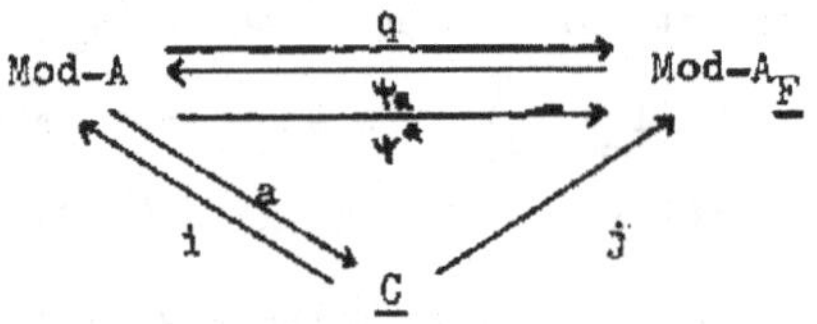

where in particular $q(M) = M_{\underline{F}}$ and $\psi^*(M) = M \otimes_A A_{\underline{F}}$. There is a natural transformation $\theta : \psi^* \to q$ where $\theta_M : M \otimes A_{\underline{F}} \to M_{\underline{F}}$ is defined as $\theta_M(x \otimes q) = \psi_M(x)q$. The diagram

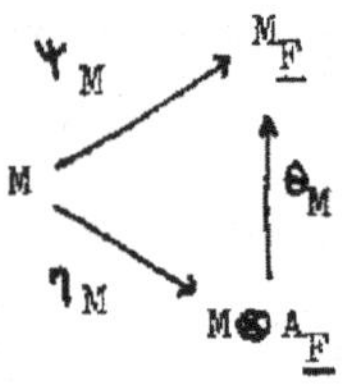

commutes.

<u>Theorem 13.1.</u> The following properties of $\underline{F}$ are equivalent:

(a) The functor $j: \underline{C} \to \text{Mod-}A_{\underline{F}}$ is an equivalence.

(b) The functor $i: \underline{C} \to \text{Mod-}A$ has a right adjoint.

(c) The functor $i: \underline{C} \to \text{Mod-}A$ is exact and commutes with direct sums.

(d) $\underline{F}$ contains a cofinal family of finitely generated right ideals and i is exact.

(e) $\theta: \psi^* \to q$ is a natural equivalence of functors.

(f) $\text{Ker}(M \to M \otimes_A A_{\underline{F}}) = t(M)$ for all M_A.

(g) $\psi(I)A_{\underline{F}} = A_{\underline{F}}$ for every $I \in \underline{F}$.

<u>Proof.</u> (a) $\Rightarrow$ (b): $i = \psi_* j$, where j has a right adjoint by hypothesis, and ψ_* has the right adjoint $\text{Hom}_A(A_{\underline{F}}, \cdot)$.

(b) $\Rightarrow$ (c): Clear, since i always is left exact.

(c) $\Leftrightarrow$ (d): Follows from Proposition 12.2.

(c) $\Rightarrow$ (e): We have $\psi_* q = ia$, which by hypothesis preserves arbitrary colimits. Since the forgetful functor ψ_* also preserves colimits, it follows that q preserves colimits. A well-known argument ([102], p.157) then shows that θ is a natural equivalence between $\cdot \otimes_A A_{\underline{F}}$ and q.

(e) $\Rightarrow$ (f): $\text{Ker}(M \to M \otimes A_{\underline{F}}) = \text{Ker}(M \to M_{\underline{F}}) = t(M)$.

(f) $\Rightarrow$ (g): If $I \in \underline{F}$, then A/I is a torsion module and hence the map $A/I \to A/I \otimes A_{\underline{F}} = A_{\underline{F}}/\psi(I)A_{\underline{F}}$ is zero. This implies $A_{\underline{F}} = \psi(I)A_{\underline{F}}$.

(g) $\Rightarrow$ (a): We must show that every right $A_{\underline{F}}$-module M is $\underline{F}$-closed as an A-module. M is torsion-free, for if $x \in M$ and $xI = 0$ for some $I \in \underline{F}$, then $xA_{\underline{F}} = xIA_{\underline{F}} = 0$ and hence $x = 0$.

We next verify that M is $\underline{F}$-injective. Let $f: I \to M$ be a homomorphism with $I \in \underline{F}$. Write $1 \in A_{\underline{F}}$ in the form $1 = \sum \psi(a_i)q_i$ with $q_i \in A_{\underline{F}}$ and $a_i \in I$. Then $J = \bigcap_i (\psi(A):q_i) \in \underline{F}$ and $I \cap J \in \underline{F}$. Since M is torsion-free, we may factor f over $\bar{f}: \psi(I) \to M$. For every $a \in I \cap J$ we get $f(a) = \bar{f}(\psi(a)) = \bar{f}(\Sigma \psi(a_i)q_i a) = \Sigma \bar{f}(\psi(a_i))q_i a = \Sigma f(a_i)q_i a$. Thus the restriction of f to $I \cap J$ has an extension to $h: A \to M$, and as usual we must then necessarily have $h|I = f$ (cf. proof of Proposition 7.8).

Definition. A topology is called perfect if it has the properties listed in the Theorem.

Corollary 13.2. If $\underline{F}$ is a perfect topology, then $A_{\underline{F}}$ is flat as a left A-module.

Proof. The functor $\cdot \otimes A_{\underline{F}} = q$ is left exact.

The main drawback of Theorem 13.1 is that none of the conditions (a) - (g) are internal, i.e. they do not give conditions for $\underline{F}$ to be perfect solely in terms of $\underline{F}$ and the ideal structure of A. So e.g. when one is applying (d), it is usually difficult to tell from $\underline{F}$ whether i is exact. In one special case this is however possible:

Proposition 13.3. If A is right hereditary, then i is exact for all topologies on A.

Proof. Exactness of i means that if $f: L \to M$ is a homomorphism between $\underline{F}$-closed modules, then also Coker f is $\underline{F}$-closed. Since Ker f is $\underline{F}$-closed, it clearly suffices to consider monomorphisms $f: L \to M$. M/L is then torsion-free, as one sees from the exact sequence

$$\mathrm{Hom}(A/I,M) \longrightarrow \mathrm{Hom}(A/I,M/L) \longrightarrow \mathrm{Ext}^1(A/I,L) .$$

Let $g: I \to M/L$ be any homomorphism with $I \in \underline{F}$. Since I is a projective module, we may lift g to $h: I \to M$. Since M is $\underline{F}$-closed, we may extend h to $A \to M$. The composition $A \to M \to M/L$ then extends g , and M/L is $\underline{F}$-injective.

Corollary 13.4. If A is a right noetherian hereditary ring, then every topology is perfect.

There is an interesting abstract characterization of those ring homomorphisms which are of the form $A \to A_{\underline{F}}$ for a perfect topology $\underline{F}$, namely they are precisely the "flat epimorphisms" in the category of rings. As a preparation for this result, we are going to review some facts about epimorphisms of rings.

Let $\varphi: A \to B$ be a ring homomorphism. Recall that φ is an epimorphism (in the category of rings) if for any ring C and homomorphisms α , $\beta: B \to C$, $\alpha\varphi = \beta\varphi$ implies $\alpha = \beta$. More generally we say that $b \in B$ is dominated by φ if $\alpha\varphi = \beta\varphi$ always implies $\alpha(b) = \beta(b)$. The set of elements of B dominated by φ is a subring of B , called the dominion of φ. φ is an epimorphism if and only if its dominion equals B .

Proposition 13.5. $\varphi: A \to B$ dominates $b \in B$ if and only if $b \otimes 1 = 1 \otimes b$ in $B \otimes_A B$.

Proof. Suppose $b \otimes 1 = 1 \otimes b$ and let α , $\beta: B \to C$ be homomorphisms such that $\alpha\varphi = \beta\varphi$. Define a homomorphism of

A-A-bimodules $\tau : B \otimes_A B \to C$ as $\tau(b \otimes b') = \alpha(b)\beta(b')$. $b \otimes 1 = 1 \otimes b$ implies $\alpha(b) = \tau(b \otimes 1) = \tau(1 \otimes b) = \beta(b)$.

Suppose on the other hand that φ dominates b . The assertion now follows by applying the following lemma to $x = 1 \otimes 1 \in B \otimes_A B$.

Lemma 13.6. Let $\varphi : A \to B$ be a ring homomorphism and M a B-B-bimodule. If $x \in M$ has the property that $\varphi(a)x = x\varphi(a)$ for all $a \in A$, then $bx = xb$ for all b in the dominion of φ .

Proof. We make $B \times M$ into a ring by defining

$$(b,y) + (b',y') = (b + b', y + y')$$

$$(b,y) \cdot (b',y') = (bb', by' + yb') .$$

The ring axioms are easily verified, in particular $B \times M$ has an identity, namely $(1,0)$. Define two maps α , $\beta : B \to B \times M$ as $\alpha(b) = (b,0)$ and $\beta(b) = (b, bx-xb)$. Both α and β are ring homomorphisms, and $\alpha\varphi = \beta\varphi$. So if b is dominated by φ , then $\alpha(b) = \beta(b)$ and thus $bx = xb$.

If one is interested in dominions in the category of commutative rings, one may prove an analogue of Proposition 13.5. The proof in the commutative case has to be done separately (for Lemma 13.6 would introduce non-commutative rings), but is easy since $B \otimes_A B$ is a ring whenever A is commutative.

We will elaborate on the consequences of Proposition 13.5. Let

$$\text{Mod-}A \underset{\varphi_*}{\overset{\varphi^*}{\rightleftarrows}} \text{Mod-}B$$

be the functors $\varphi^*(M) = M \otimes_A B$, $\varphi_*(N) = N$. Recall that φ^* is a left adjoint of φ_* .

Proposition 13.7. The following properties of a ring homomorphism $\varphi : A \to B$ are equivalent:

(a) φ is an epimorphism.

(b) The canonical map $B \otimes_A B \to B$ is bijective.

(c) The adjunction transformation $\varphi^* \varphi_* \to \mathrm{id.}$ is a natural equivalence of functors.

(d) The functor $\varphi_* : \text{Mod-}B \to \text{Mod-}A$ is full.

Proof. (a) $\Rightarrow$ (d): Suppose M and N are B-modules and $\alpha : M \to N$ is A-linear. For each $x \in M$ consider the map $\beta : B \otimes_A B \to N$ given by $\beta(b \otimes b') = \alpha(xb)b'$. Note that this really is a welldefined map. Since $1 \otimes b = b \otimes 1$, we have $\alpha(xb) = \alpha(x)b$, and therefore α is B-linear.

(d) $\Rightarrow$ (c): We have to show that the homomorphism $\mu : M \otimes_A B \to M$ given by $x \otimes b \mapsto xb$, is an isomorphism for every B-module M. The map $M \to M \otimes_A B$ given by $x \mapsto x \otimes 1$ is clearly A-linear, and by hypothesis therefore B-linear. It is the desired inverse of μ.

(c) $\Rightarrow$ (b) is clear, while (b) $\Rightarrow$ (a) follows from Proposition 13.5.

An immediate consequence of (d) is:

Corollary 13.8. Let $\varphi : A \to B$ be an epimorphism. If M is a right B-module such that M_A is injective, then M_B is injective.

For the proof of the next theorem we need the following result:

<u>Lemma 13.9.</u> Let there be given modules L_A and ${}_AM$. The following assertions are equivalent:

(a) $xA \otimes M = 0$ for every $x \in L$.

(b) For every $y \in M$ and $x_1, \dots, x_n \in L$, there exist $y_1, \dots, y_m \in M$ and $a_1, \dots, a_m \in A$ such that $y = \sum_1^m a_i y_i$ and $x_i a_j = 0$ for all i and j.

(c) For every $y \in M$ and $x \in L$, there exist $y_1, \dots, y_m \in M$ and $a_1, \dots, a_m \in A$ such that $y = \sum_1^m a_i y_i$ and $xa_j = 0$ for all j.

<u>Proof.</u> The equivalence of (a) and (c) is well-known ([10], ch. 1, § 2, Lemma 10). (a) implies that $C \otimes M = 0$ for every submodule C of L, and one easily proves that this implies $C \otimes M = 0$ for every submodule C of a direct sum L^n of copies of L. (b) is then obtained from (a) by considering $(x_1, \dots, x_n)$ as an element of L^n.

<u>Theorem 13.10.</u> Let $\varphi : A \to B$ be a ring homomorphism. The following assertions are equivalent:

(a) φ is an epimorphism and makes B into a flat left A-module.

(b) The family $\underline{F}$ of right ideals I of A such that $\varphi(I)B = B$ is a topology, and there exists a ring isomorphism $\sigma : B \to A_{\underline{F}}$ such that $\sigma\varphi = \psi_A$.

(c) The following two conditions are satisfied:

(i) for every $b \in B$ there exist $s_1, \dots, s_n \in A$ and $b_1, \dots, b_n \in B$ such that $b\varphi(s_i) \in \varphi(A)$ and $\Sigma\varphi(s_i)b_i = 1$;

(ii) if $\varphi(a) = 0$, then there exist $s_1 \dots, s_n \in A$ and $b_1, \dots, b_n \in B$ such that $as_i = 0$ and $\Sigma\varphi(s_i)b_i = 1$.

<u>Proof.</u> (a) $\Rightarrow$ (b): The forgetful functor φ_*: Mod-B $\to$ Mod-A makes Mod-B equivalent to a full subcategory of Mod-A by Proposition 13.7(d). φ_* has a left adjoint $\varphi^* = \cdot \otimes_A B$ which is exact, since ${}_A B$ is flat. We may thus consider Mod-B as a Giraud subcategory of Mod-A . From Theorem 10.2, combined with Theorem 13.1, we may conclude that the family $\underline{F}$ of right ideals I such that $\varphi^*(I) \cong \varphi^*(A)$, i.e. such that $\varphi(I)B = B$, is a topology and that $A_{\underline{F}} \cong B$.

(b) $\Rightarrow$ (c): If $b \in B = A_{\underline{F}}$, there exists $I \in \underline{F}$ such that $bI \subset \varphi(A)$. By Theorem 13.1 we may assume I finitely generated, say generated by $s_1, \ldots, s_n$. Then $1 = \sum \varphi(s_i)b_i$ for some $b_i \in B$, and (i) is verified.

If $\varphi(a) = 0$, then $a \in \mathrm{Ker}(A \to A_{\underline{F}})$, so there exists $I \in \underline{F}$ such that $aI = 0$. As before one may assume I generated by $s_1, \ldots, s_n$, and $1 = \sum \varphi(s_i)b_i$. Thus also (ii) is verified.

(c) $\Rightarrow$ (a): First of all we note:

<u>Lemma 13.11.</u> If $\varphi: A \to B$ is a ring homomorphism satisfying condition (i) of (c), then $C \otimes_A B = 0$ for every submodule C of $B/\varphi(A)$

<u>Proof.</u> (i) implies that 13.9(c) is satisfied.

We now prove (c) $\Rightarrow$ (a). Since we have just seen that $B/\varphi(A) \otimes_A B = 0$, we obtain epimorphisms

$$B \cong A \otimes_A B \longrightarrow \varphi(A) \otimes_A B \longrightarrow B \otimes_A B \quad .$$

The canonical map $B \otimes_A B \to B$ must then be a bijection, and this shows that φ is an epimorphism (Proposition 13.7).

It remains to see ${}_AB$ is flat. For this we use Proposition 13 of [10], ch. 1, § 2. Suppose we have $s_1,\ldots,s_n \in A$ and $b_1,\ldots,b_n \in B$ such that $\Sigma\varphi(s_i)b_i = 0$. Applying Lemma 13.9(b) to $1 \in B$ and $\bar{b}_1,\ldots,\bar{b}_n \in B/\varphi(A)$, we obtain $a_1,\ldots,a_m \in A$ and $b'_1,\ldots,b'_m \in B$ such that

$$1 = \sum_j \varphi(a_j)b'_j$$

$$b_i\varphi(a_j) = \varphi(c_{ij}) \quad \text{for some} \quad c_{ij} \in A .$$

Then $\sum_i \varphi(s_i c_{ij}) = \sum_i \varphi(s_i)b_i\varphi(a_j) = 0$, and so $\sum_i s_i c_{ij} \in \operatorname{Ker}\varphi$ for each j. We now make use of condition (ii), and obtain for each j elements $t_{j1},\ldots,t_{jr} \in A$ and $b''_{j1},\ldots,b''_{jr} \in B$ such that

$$\sum_i s_i c_{ij} t_{jk} = 0$$

$$\sum_k \varphi(t_{jk})b''_{jk} = 1 .$$

We then have

$$b_i = \sum_j b_i\varphi(a_j)b'_j = \sum_j \varphi(c_{ij})b'_j = \sum_{j,k} \varphi(c_{ij}t_{jk})b''_{jk}b'_j$$

and $\sum_i s_i c_{ij} t_{jk} = 0$, i.e. the given relation comes from a relation in A, as was to be shown.

<u>Remark 1:</u> Note that condition (c) of the theorem implies in particular that each $b \in B$ may be written as $b = \Sigma\, b\varphi(s_i)b_i = \Sigma\varphi(a_i)b_i$ with $\Sigma\varphi(s_i)b_i = 1$. B is thus obtained by a sort of generalized calculus of fractions (cf. §15).

Remark 2: As was noted in the proof of (c) ⇒ (a), condition (i) of (c) may be restated as:

(i)' For every family $b_1,\dots,b_r \in B$ there exist $s_1,\dots,s_n \in A$ and $b'_1,\dots,b'_n \in B$ such that $b_j\varphi(s_i) \in \varphi(A)$ and $\sum_i \varphi(s_i)b'_i = 1$.

Corollary 13.12. There is a 1-1 correspondence between perfect topologies on A and equivalence classes of ring epimorphisms $A \to B$ such that ${}_AB$ is flat.

Exercises:

1. Show that a topology $\underline{F}$ is perfect if and only if all $A_{\underline{F}}$-modules are torsion-free as A-modules.
2. Suppose $\underline{F}$ is a perfect topology. Show that:
   (i) M_A is a torsion module if and only if $M \otimes_A A_{\underline{F}} = 0$.
   (ii) $t(M) = Tor_1(M, A_{\underline{F}}/A)$ for all modules M, assuming A torsion-free.
3. Let $A \subset B$ be commutative rings. Show that $A \hookrightarrow B$ is a flat epimorphism if and only if $(A:b)B = B$ for every $b \in B$. (Hint: note that if $I_\alpha B = B$ for a finite family of ideals I_α, then $(\bigcap I_\alpha)B = B$).

References: Goldman [33], Lambek [48] (§ 2), Popescu and Spircu [104], Roos [66] (ch. 1), Silver [70], Walker and Walker [81].

For commutative rings: Akiba [84], Năstăsescu and Popescu [59], Sém. Samuel [67] (Exposé 6 by Olivier).

§ 14. Maximal flat epimorphic extension of a ring

Theorem 14.1. For every ring A there exists a ring $M(A)$ and a ring homomorphism $\varphi: A \to M(A)$ such that

(i) φ is an injective epimorphism and $M(A)$ is a flat left A-module.

(ii) For every injective epimorphism $\alpha: A \to B$ of rings such that ${}_A B$ is flat, there exists a unique ring homomorphism $\beta: B \to M(A)$ such that $\beta\alpha = \varphi$; moreover, β is also injective.

Proof. Every flat epimorphism is obtained as the canonical homomorphism $\psi: A \to A_{\underline{F}}$ for a perfect topology $\underline{F}$ (Theorem 13.10). ψ is injective if and only if A has no $\underline{F}$-torsion, i.e. if and only if $\underline{F} \subset \underline{D}$ (the family of dense right ideals). So if $M(A)$ exists, it should be a subring of the maximal ring of quotients $A_{\underline{D}} = Q_m$. This leads us to consider the family $\underline{P}$ of all subrings B of Q_m such that $A \subset B$ and $A \hookrightarrow B$ is a flat epimorphism. Note that inclusion of subrings in $\underline{P}$ corresponds to inclusion of the corresponding perfect topologies.

Lemma 14.2. The family $\underline{P}$ is directed under inclusion.

Proof. We will show that if B and C are members of $\underline{P}$, then the smallest subring D of Q_m containing both B and C is also a member of $\underline{P}$. Every element of D is a sum of elements of the form

$(*)$ $d = b_1c_1 \ldots . b_nc_n$ with $b_i \in B$, $c_i \in C$, and we may assume that each d appearing in a given sum has the same length n. We will verify condition (i) of 13.10(c), i.e.:

given $d_1,\dots,d_r$ of length n, there exist $d_1',\dots,d_m'$ in D and $s_1,\dots,s_m \in A$ such that $d_i s_k \in A$ for all i, k and $\sum s_k d_k' = 1$.

We do this by induction on the length n. For $n = 0$, i.e. $d = 1$, the condition is clearly satisfied. Suppose $n > 1$ and the condition has been verified for $n-1$. To simplify the notation somewhat, we only consider the case $r = 1$; it is easy to see that our argument extends immediately to the case of a family $d_1,\dots,d_r \in D$. So suppose we are given d of the form (*). By the induction hypothesis there exist $d_1',\dots,d_m'$ in D and $t_1,\dots,t_m$ in A such that

$x_i = b_2 c_2 \dots b_n c_n t_i \in A$ for all i, and $\sum t_i d_i' = 1$.

By the remark 2 of § 13, applied to C, there exist $c_1',\dots,c_p'$ in C and $s_1,\dots,s_p$ in A such that

$x_{ij} = c_1 x_i s_j \in A$ for all i, j, and $\sum s_j c_j' = 1$.

Similarly there exist $b_1',\dots,b_q'$ in B and $r_1,\dots,r_q$ in A such that

$b_1 x_{ij} r_k \in A$ for all i, j, k, and $\sum r_k b_k' = 1$.

We have thus got elements $b_k' c_j' d_i' \in D$ and $t_i s_j r_k \in A$ such that

$d t_i s_j r_k = b_1 c_1 x_i s_j r_k \in A$ and $\sum\limits_{i,j,k} t_i s_j r_k b_k' c_j' d_i' = 1$.

This finishes the proof of the Lemma, and we may continue the proof of the Theorem. Define $M(A)$ as the union of all rings in $\underline{P}$. It is obvious that $\varphi: A \to M(A)$ is an epimorphism, and $M(A)$ is flat as a left A-module since it is a direct limit of flat A-modules.

Suppose $\alpha: A \to B$ is any other injective flat epimorphism. There is a corresponding perfect topology $\underline{F}$, and as we noticed above, we have $\underline{F} \subset \underline{D}$ and hence a commutative diagram

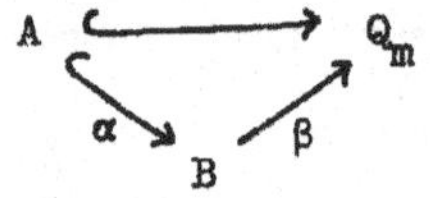

But also $\beta: B \to Q_m$ must be injective, because $\beta(b) = 0$ would mean that there exists a homomorphism $f: I \to A$ with $I \in \underline{F}$ such that $f|J = 0$ for some $J \in \underline{D}$; for each $a \in I$ we have $(J:a) \in \underline{D}$ and is $\underline{D}$-torsion-free, so it follows that $f = 0$. We conclude that the image of β lies within $M(A)$. Since β obviously is unique (α is an epimorphism), we have proved the Theorem.

We will investigate the properties of $M(A)$.

<u>Proposition 14.3.</u> If $f: A \to B$ is a ring epimorphism such that ${}_A B$ is flat, then $J = f^{-1}(J)B$ for every right ideal J of B .

<u>Proof.</u> Put $I = f^{-1}(J)$. Tensoring the inclusion $A/I \hookrightarrow B/J$ by B , we obtain $B/IB \hookrightarrow B/J \otimes_A B = B/J$ by 13.7(c). Since $B/IB \to B/J$ obviously is surjective, it is an isomorphism and $J = IB$.

<u>Corollary 14.4.</u> If A is right noetherian, then so is also $M(A)$.

<u>Proposition 14.5.</u> If A is von Neumann regular, then $M(A) = A$.

<u>Proof.</u> By Lemma 13.11 we have $M(A)/A \otimes_A M(A) = 0$. But since

A is regular, $A \hookrightarrow M(A)$ induces a monomorphism $M(A)/A \otimes_A A \hookrightarrow M(A)/A \otimes_A M(A)$, so $M(A)/A = 0$.

Further results on $M(A)$ may be obtained by using homological methods. When $A \to B$ is a ring homomorphism such that ${}_AB$ is flat, one has the following formulas ([13], ch. 6, § 4):

(1) $\quad \mathrm{Ext}_A^n(M,N) \cong \mathrm{Ext}_B^n(M \otimes_A B, N)$ for M_A , N_B ;

(2) $\quad \mathrm{Tor}_n^A(M,N) \cong \mathrm{Tor}_n^B(M \otimes_A B, N)$ for M_A , ${}_BN$.

Combining these formulas with 13.7(c) we obtain:

<u>Proposition 14.6.</u> If $\varphi : A \to B$ is a ring epimorphism such that ${}_AB$ is flat, then

$\mathrm{Ext}_A^n(M,N) \cong \mathrm{Ext}_B^n(M,N)$ for M_B , N_B ;

$\mathrm{Tor}_n^A(M,N) \cong \mathrm{Tor}_n^B(M,N)$ for M_B , ${}_BN$.

<u>Corollary 14.7.</u> $\text{r.gl.dim } M(A) \leq \text{r.gl.dim } A$ and

$\text{w.gl.dim } M(A) \leq \text{w.gl.dim } A$,

where r.gl.dim is the right global dimension and w.gl.dim is the weak global dimension.

The ring A is called <u>right coherent</u> if every direct product of flat left A-modules is flat [17].

<u>Proposition 14.8.</u> If A is right coherent, then so is also $M(A)$.

<u>Proof.</u> Let (N_α) be a family of flat left $M(A)$-modules. Then each N_α is flat as an A-module by (2). Thus $\prod_\alpha N_\alpha$ is flat over A , but is then flat also over $M(A)$ by 14.6.

Example:

It follows from Corollary 13.4 that when A is right noetherian hereditary, then $M(A) = Q_m$. This result will be improved in § 20 (Theorem 20.2).

Exercises:

1. Let $\underline{F}$ be the Goldie topology (§ 3, Example 1). Show that if $\underline{F}$ is perfect, then $A_{\underline{F}}$ is right noetherian.
2. Suppose A is a ring for which Q_{cl} exists and such that every finitely generated right ideal of A is principal. Show that $Q_{cl} = M(A)$.

References: Knight [99], Popescu and Spircu [104]. For the commutative case: Akiba [84], Lazard [100] (ch. 4).

§ 15. 1-topologies and rings of fractions

A 1-topology on A is a topology containing a cofinal family of principal right ideals. A 1-topology $\underline{F}$ is determined by the set $\Sigma(\underline{F}) = \{ s \in A \mid sA \in \underline{F} \}$.

Proposition 15.1. The map Σ defines a 1-1 correspondence between 1-topologies on A and subsets S of A satisfying:

S1. $1 \in S$.

S2. $s, t \in S$ implies $st \in S$.

S3. If $s \in S$ and $a \in A$, then there exist $t \in S$ and $b \in A$ such that $sb = at$ (S is right permutable).

S4. If $ab \in S$, then $a \in S$.

<u>Proof.</u> Let $\underline{F}$ be a 1-topology. $\Sigma(\underline{F})$ obviously satisfies S 1. S 2 follows from axiom T 2 for topologies, because for each $sa \in sA$ we have $(stA:sa) \supset (tA:a) \in \underline{F}$ by T 1. S 3 is clear from T 1, because we have $(sA:a) \supset tA$ for some $tA \in \underline{F}$. S 4 is immediate from T 3.

Conversely, if S satisfies S 1-3 and one sets $\underline{F} = \{ I \mid I \supset sA$ for some $s \in S\}$, then $\underline{F}$ is easily verified to be a 1-topology. S 4 is a saturation axiom which makes the correspondence $\underline{F} \leftrightarrow \underline{S}$ one-to-one.

For a 1-topology one may describe the modules of quotients in a rather explicit way.

<u>Proposition 15.2.</u> If $\underline{F}$ is a 1-topology and $M \in$ Mod-A, then

$$M_{\underline{F}} = \{ (x,s) \in M \times S \mid sa = 0 \text{ in } A \text{ implies } xat = 0 \text{ for some } t \in S\}/\sim$$

where $\sim$ is the equivalence relation given by $(x,s) \sim (y,t)$ if there exist $a, b \in A$ such that $sa = tb \in S$ and $xa = yb$.

<u>Proof.</u> Recall that we have $M_{\underline{F}} = \varinjlim \mathrm{Hom}_A(sA, M/t(M))$ with $s \in S$. A homomorphism $\varphi: sA \longrightarrow M/t(M)$ is dtermined by an element $x \in M$ such that $sa = sb$ in A implies $xa - xb \in t(M)$, i.e. $xat = xbt$ for some $t \in S$. In the limit, φ gives the same element of $M_{\underline{F}}$ as $\psi: tA \longrightarrow M/t(M)$, determined by $y \in M$, if and only if φ and ψ coincide on some $uA \subset sA \cap tA$ with $u \in S$, i.e. if and only if there exist $a, b \in A$ such that $u = sa = tb \in S$ and $xa - yb \in t(M)$. This clearly corresponds to the relation $\sim$.

One easily verifies that under the isomorphism described in the Proposition, the module operations in $M_{\underline{F}}$ take the form:

$(x,s) + (y,t) = (xa+yb,u)$ where $sa = tb = u \in S$;

$(x,s) \cdot (a,t) = (xb,tv)$ for some $b \in A$, $v \in S$ such that

$$av = sb .$$

<u>Proposition 15.3.</u> A 1-topology $\underline{F}$ is perfect if and only if $S = \Sigma(\underline{F})$ satisfies:

S5'. For every $s \in S$ there exists $a \in A$ such that $sa \in S$ and such that $sab = 0$ implies $abu = 0$ for some $u \in S$.

<u>Proof.</u> Perfectness of $\underline{F}$ means that for every $s \in S$ there exists $q \in A_{\underline{F}}$ such that $sq = 1$ (Theorem 13.1(g)). Suppose that so is the case. Using 15.2, we write $q = (a,t)$. Then $(s,1)(a,t) = (sa,t) \sim (1,1)$, so there exist a', $b' \in A$ such that $ta' = 1 \cdot b' \in S$ and $saa' = 1 \cdot b'$. We may then take aa' as the element a in S 5', for $saa' = b' \in S$ and if $saa'b = 0$, then $ta'b = b'b = saa'b = 0$ and $(a,t) \in A_{\underline{F}}$ implies $aa'bu = 0$ for some $u \in S$ (by 15.2).

Conversely, if for each $s \in S$ there exists $a \in A$ such that as in S 5' , then (a,sa) represents an element in $A_{\underline{F}}$ and one has $(s,1)(a,sa) = 1$.

The axiom S 5' is a weakened form of the perhaps more well-known condition:

S5. If $sa = 0$ with $s \in S$, then $at = 0$ for some $t \in S$ (S is <u>right reversible</u>).

The most important examples of 1-topologies are those given by the rings of fractions. Let S be a multiplicatively closed subset of A . A <u>right ring of fractions</u> of A with respect to S is a ring $A[S^{-1}]$ and a ring homomorphism $\varphi : A \to A[S^{-1}]$ satisfying:

F1. $\varphi(s)$ is invertible for every $s \in S$.

F2. every element in $A[S^{-1}]$ has the form $\varphi(a)\varphi(s)^{-1}$ with $s \in S$.

F3. $\varphi(a) = 0$ if and only if $as = 0$ for some $s \in S$.

Similarly one defines the <u>left ring of fractions</u> $[S^{-1}]A$ of A with respect to S . It is not immediately clear that the axioms F 1-3 dtermine $A[S^{-1}]$ uniquely, but that so is the case follows from the fact $A[S^{-1}]$ is a solution of a universal problem:

<u>Proposition 15.4.</u> If $A[S^{-1}]$ exists, it has the following property: for every ring homomorphism $\psi : A \to B$ such that $\psi(s)$ is invertible in B for every $s \in S$, there exists a unique homomorphism $\sigma : B \to A[S^{-1}]$ such that $\sigma\varphi = \psi$.

<u>Proof.</u> We define σ as $\sigma(\varphi(a)\varphi(s)^{-1}) = \psi(a)\psi(s)^{-1}$. We then have to verify that this is well-defined. So suppose $\varphi(a)\varphi(s)^{-1} = \varphi(b)\varphi(t)^{-1}$. Then $\varphi(a) = \varphi(b)\varphi(t)^{-1}\varphi(s) =$ $= \varphi(b)\varphi(c)\varphi(u)^{-1}$ for some $c \in A$, $u \in S$, by F 2. So $\varphi(a)\varphi(u) = \varphi(b)\varphi(c)$, and by F 3 this implies that $auv =$ $= bcv$ for some $v \in S$. Then $\psi(a)\psi(u) = \psi(b)\psi(c)$ since $\psi(v)$ is invertible, and we may go backwards to obtain

$\psi(a)\psi(s)^{-1} = \psi(b)\psi(t)^{-1}$. We leave to the reader to verify that σ is a homomorphism. It is clear that $\sigma\varphi = \psi$ and that σ is unique.

Corollary 15.5. $A[S^{-1}]$ is unique up to isomorphism.

The unicity of the solution of a universal problem also implies:

Corollary 15.6. If both $A[S^{-1}]$ and $[S^{-1}]A$ exist, then they are isomorphic.

Since there are examples of A , S such that $[S^{-1}]A$ exists but $A[S^{-1}]$ does not exist ([10], p. 163), a ring may satisfy the universal property of 15.4 without being a right ring of fractions with respect to S . We now turn to the question of the existence of $A[S^{-1}]$.

Proposition 15.7. Let S be multiplicatively closed in A . $A[S^{-1}]$ exists if and only if S is both right permutable and right reversible. If $A[S^{-1}]$ exists, then $A[S^{-1}] = A_{\underline{F}}$ where the topology $\underline{F} = \{I \mid I \supset sA$ for some $s \in S\}$ is perfect.

Proof. If S is right permutable and right reversible, then $\underline{F}$ is a perfect topology by Proposition 15.3. Let $\psi: A \to A_{\underline{F}}$ be the canonical homomorphism, and use the description of $A_{\underline{F}}$ in Proposition 14.2. S 5 guarantees that $(1,s)$ represents an element of $A_{\underline{F}}$ whenever $s \in S$. $(1,s)$ will be an inverse of $\psi(s)$. Every element $(a,s) \in A_{\underline{F}}$ has the form $(a,1)(1,s) = \psi(a)\psi(s)^{-1}$. Finally, if $(a,1) \sim (0,1)$, then $as = 0$ for some $s \in S$. Hence ψ satisfies F 1-3.

Suppose conversely $A[S^{-1}]$ exists. S is then right permutable, for if $a \in A$, $s \in S$ are given, then $\varphi(s)^{-1}\varphi(a) = \varphi(b)\varphi(t)^{-1}$ by F 2, i.e. $\varphi(at) = \varphi(sb)$. By F 3 this means that $atu = sbu$ for some $u \in S$. Since $tu \in S$, we have S 3. If $sa = 0$ with $s \in S$, then $\varphi(a) = 0$ by F 1 and $at = 0$ for some $t \in S$ by F 3, so we have S 5.

<u>Corollary 15.8.</u> There is a 1-1 correspondence between right rings of fractions of A and subsets of A satisfying S 1-5.

For $A[S^{-1}]$ one may simplify the formula of Proposition 15.2 somewhat, in that

$M_{\underline{F}} = M \times S/\sim$, where $\sim$ is defined as before.

Of course one has $M_{\underline{F}} = M \otimes_A A[S^{-1}]$.

<u>Examples:</u>

1. When A is commutative, S 3 and S 5 are automatically satisfied. The theory of rings of fractions is well-known in that case [10].

2. Let S be the set of non-zero-divisors of A . $A[S^{-1}]$ is called the <u>classical right ring of quotients</u> of A , and will be denoted by Q_{cl} . It is a subring of the maximal right ring of quotients Q_m of A , and is in fact a subring of M(A) . From Proposition 15.7 we get:

<u>Proposition 15.9.</u> The classical right ring of quotients of A exists if and only if A satisfies the "Ore condition", i.e. for $a \in A$ and a non-zero-divisor s there exist $b \in A$ and a non-zero-divisor t such that $sb = at$.

Note that if A has both a classical right ring of quotients and a classical left ring of quotients, then these two rings coincide, by Corollary 15.6.

<u>Exercises</u>:

1. Show that if A has no nilpotent elements $\neq 0$, then S 5 is always satisfied.
2. Let $A[S^{-1}]$ be a ring of fractions. Show that an A-module M is $\underline{F}$-closed for the corresponding topology $\underline{F}$ if and only if M is torsion-free and divisible (i.e. $M = Ms$ for every $s \in S$).
3. Let A be a regular ring and let $S = \{a \in A \mid ba = 0$ implies $b = 0\}$. Show that S satisfies S 1-5' and that S satisfies S 5 only if all elements in S are invertible. (Illustration: A is the endomorphism ring of an infinite-dimensional vector space).

<u>References</u>: Almkvist [3], Bourbaki [10] (p. 162-163), Eriksson [26], Gabriel [31], Gabriel-Zisman [89].

Chapter 4. Self-injective rings

§ 16. The endomorphism ring of an injective module

A ring A is called regular (in the sense of von Neumann) if for every $a \in A$ there exists $x \in A$ such that $axa = a$. We recall the following alternative characterizations of regular rings (cf. [10], p. 64):

Proposition 16.1. The following properties of A are equivalent:

(a) A is regular.

(b) Every finitely generated right ideal of A is generated by an idempotent element.

(c) Every right A-module is flat.

It follows from (b) that every right noetherian regular ring is semi-simple. More generally, (b) implies that if A is regular and has no infinite family of orthogonal idempotents, then A is semi-simple.

Theorem 16.2. Let E be an injective A-module with endomorphism ring H and let J be the Jacobson radical of H. Then:

(i) H/J is regular.

(ii) Idempotents may be lifted modulo J.

The last assertion means that if e is an idempotent in H/J, then there exists an idempotent in H mapping canonically onto e. It is well-known that if idempotents may be lifted, then one may lift any countable orthogonal family of idempotents in H/J so that orthogonality is preserved ([47], § 3.6). The proof of the Theorem will be broken up into several steps. We define

$N = \{ h \in H \mid \text{Ker } h \text{ is an essential submodule of } E \}$.

<u>Lemma 16.3.</u> N is a two-sided ideal of H, and H/N is a regular ring.

<u>Proof.</u> If $f, g \in N$, then $f+g \in N$ since $\text{Ker}(f+g) \supset \text{Ker } f \cap \text{Ker } g$. If $f \in N$ and $h \in H$, then $fh \in N$ since $\text{Ker } fh = h^{-1}(\text{Ker } f)$, and $hf \in N$ since $\text{Ker } hf \supset \text{Ker } f$. N is thus a two-sided ideal. It remains to show regularity of H/N. Let $h \in H$. Choose a submodule K of E which is maximal with the property that $K \cap \text{Ker } h = 0$. $K + \text{Ker } h$ is then an essential submodule of E. Since the restriction of h to K is a monomorphism, and since E is injective, there exists $g: E \to E$ such that $gh(x) = x$ for $x \in K$.

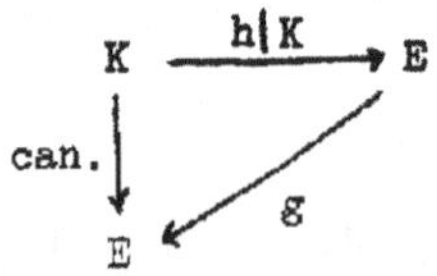

If $y \in K + \text{Ker } h$, we write $y = x+z$ with $x \in K$ and $h(z) = 0$. Then $hgh(y) = hgh(x) = h(x) = h(y)$, so $hgh-h \in N$. This shows that H/N is regular.

<u>Lemma 16.4.</u> $N = J$.

<u>Proof.</u> Suppose $h \in N$. Since $\text{Ker}(1-h) \cap \text{Ker } h = 0$ and $\text{Ker } h$

is essential in E, it follows that $1-h$ is a monomorphism. $\mathrm{Im}(1-h)$ must then be a direct summand of E, because E is injective. If $h(x) = 0$, then $x = (1-h)(x)$, so $\mathrm{Im}(1-h) \supset$ $\supset \mathrm{Ker}\, h$. $\mathrm{Im}(1-h)$ is therefore essential in E, and we must have $\mathrm{Im}(1-h) = E$. $1-h$ is thus invertible for every $h \in N$, and so $N \subset J$.

On the other hand, since the Jacobson radical is defined as the intersection of all maximal right ideals, it is clear that the radical of H/N is J/N. But H/N is regular by 16.3, so $J/N = 0$ and $J = N$.

To conclude the proof of the Theorem, it remains to show that idempotents may be lifted mod J. Assume $h \in H$ and $h-h^2 \in J$, which means that $L = \mathrm{Ker}(h-h^2)$ is essential in E. The injective envelope of $h(L)$ is a direct summand of E and thus has the form $\mathrm{Im}\, e$ for some idempotent e in H. Then $eh = h$ on L, so $eh-h \in N$. Put $f = e + eh(1-e)$ and note that $f = f^2$. Put $L' = \mathrm{Im}(1-e) + h(L)$ and note that L' is an essential submodule of $\mathrm{Im}(1-e) + \mathrm{Im}\, e = E$. $f-eh = e-ehe$, and therefore $f = eh$ on L', so $f-eh \in N$. Since we already have $eh-h \in N$, it follows that $f-h \in N = J$.

<u>Corollary 16.5.</u> If A is a right self-injective ring with Jacobson radical J, then:

(i) J is the right singular ideal of A.

(ii) A/J is regular.

(iii) Idempotents may be lifted modulo J.

We remark that the ring H/J of the Theorem may be shown to be right self-injective (Osofsky [61], Renault [107], Roos [109]), but this is a fact which we will not need. Instead we are interested in the case when H/J is semi-simple.

Definition. The ring A is semi-perfect if A/J is semi-simple, where J is the Jacobson radical of A, and idempotents may be lifted mod J.

It is well-known that one can always lift idempotents when J is a nilideal ([47], § 3.6), and therefore every right or left artinian ring is semi-perfect. From 16.5 follows immediately:

Proposition 16.6. A right self-injective ring is semi-perfect if and only if it has no infinite family of orthogonal idempotents.

A module M is called finite-dimensional if it does not contain any infinite family of non-zero submodules M_i such that their sum ΣM_i is direct. It is clear that an injective module is finite-dimensional if and only if its endomorphism ring has no infinite family of orthogonal idempotents. Hence Theorem 16.2 gives the following generalization of 16.6:

Proposition 16.7. The endomorphism ring of an injective module E is semi-perfect if and only if E is finite-dimensional.

Examples:

1. If E is an indecomposable injective module, then H/J is a division ring ([52]).

2. The ring A is called <u>right finite-dimensional</u> if A_A is a finite-dimensional module (cf. § 11, Exercise 4). Such a ring cannot have any infinite family of orthogonal idempotents.

Exercises:

1. Show that a commutative ring is semi-perfect if and only if it is a product of finitely many local rings.
2. Show that a semi-perfect ring is a direct sum of indecomposable right ideals.
3. Show that a module M is finite-dimensional if and only if E(M) is finite-dimensional.

References: Faith [28], § 5 , Lambek [47], § 4.4.

§ 17. Coperfect rings

As a natural generalization of the class of artinian rings we define:

Definition. A is a semi-primary ring if the Jacobson radical J is nilpotent and A/J is semi-simple.

If A is semi-primary and right noetherian, then A is right artinian, by a classical argument ([9], § 6, Prop. 12).

A further generalization leads to:

Definition. A is a <u>right coperfect</u> ring if it satisfies DCC on finitely generated right ideals.

The usual Zorn's lemma argument shows that the DCC is equivalent to the condition that every non-empty family of finitely generated right ideals contains a minimal member. It is known that a ring is right coperfect if and only if it is left perfect in the sense of Bass [7] (Björk [86]), but this fact we do not have to use.

<u>Proposition 17.1.</u> Every semi-primary ring is right and left coperfect.

<u>Proof.</u> We use induction on the smallest integer $n \geq 1$ such that $J^n = 0$. When $n = 1$, A is semi-simple and obviously right and left coperfect. Suppose the assertion has been proved for all semi-primary rings such that $J^m = 0$ for some $m < n$. Let A be semi-primary with $J^n = 0$ but $J^{n-1} \neq 0$. Suppose $I_1 \supset I_2 \supset \ldots$ is a descending chain of finitely generated right ideals of A. The canonical map $A \to A/J^{n-1}$ takes this chain into a chain $\bar{I}_1 \supset \bar{I}_2 \supset \ldots$ of finitely generated right ideals of A/J^{n-1}. But the radical of A/J^{n-1} is J/J^{n-1}, so this later chain is stationary by the induction hypothesis. Hence we have $I_k \subset I_r + J^{n-1}$ for all $r > k$, for some k. Since $J^n = 0$, this gives $I_kJ = I_rJ$ for all $r > k$. Now consider the chain $I_1/I_kJ \supset I_2/I_kJ \supset \ldots$ of right ideals over the semi-simple ring A/J. This chain is stationary, so for some s we have $I_s \subset I_t + I_kJ = I_t + I_sJ$ (all $t \geq s$) since we may assume $s \geq k$. The Nakayama lemma ([9], § 6, Cor. 2 of Prop. 6) gives $I_s = I_t$ for $t \geq s$.

Some basic properties of coperfect rings are summarized in the following statement:

Proposition 17.2. If A is right coperfect, then:

(i) A is right semi-artinian.

(ii) J is a nilideal.

(iii) A is semi-perfect.

Proof. (i): It suffices to show that every cyclic right module A/I contains a simple submodule. Let I' be a minimal finitely generated right ideal $\not\subset$ I . I+I'/I is then a simple submodule of A/I .

(ii): Let s be the preradical associating to each module its socle. From § 3, Example 5, we recall that $J = \bar{s}(J)$, i.e. $J = s_\beta(J)$ for some ordinal β . For each $a \in J$ we define o(a) as the smallest β such that $a \in s_\beta(J)$. o(a) is never a limit ordinal, for if $a \in \bigcup_{\alpha<\beta} s_\alpha(J)$, then $a \in s_\alpha(J)$ for some α . Hence we may write $o(a) = \alpha+1$ for some ordinal α . Recall that $s_{\alpha+1}(J)/s_\alpha(J) = s(J/s_\alpha(J))$ and is therefore annihilated by J , so $s_{\alpha+1}(J)\cdot J \subset s_\alpha(J)$. This shows that for any $a, b \in J$ we have $h(ab) < h(a)$. If $a \in J$ were not nilpotent, the sequence $h(a^n)$ would be an infinite strictly decreasing sequence of ordinals, which is impossible.

(iii): Since J is a nilideal, we may lift idempotents. It remains to see that A/J is semi-simple. It is easy to see that A/J also is right semi-artinian. If I is a minimal right ideal of A/J , then there exists a maximal ideal I' such that $I \cap I' = 0$ since the Jacobson radical of A/J is zero. I is therefore generated by an idempotent. Since we can lift idempotents modulo J , and since a right coperfect ring obviously cannot have any infinite family of orthogonal idempotents, the right socle of A/J is a finite direct sum of minimal right ideals, and is therefore itself a direct summand of A/J . Its complementary summand must then have zero socle and is therefore zero. Hence A/J is semi-simple.

It should be remarked here that it can be shown that conditions (i) and (iii) of 17.2 imply conversely that A is left perfect (Bass [7]) and hence right coperfect (Björk [86]).

Proposition 17.3. If A is right noetherian and right or left coperfect, then A is right artinian.

Proof. If A is right noetherian and right coperfect, then A is obviously right artinian. Suppose A is right noetherian and left coperfect. J is a nilideal, but every nilideal in a noetherian ring is nilpotent ([47], p. 70). Since A/J is semi-simple by 17.2, A is a semi-primary ring and hence right artinian by a previous remark.

This result may be generalized somewhat. For this we introduce some terminology. For each subset S of A we put

$$r(S) = \{ a \in A \mid Sa = 0 \} , \quad l(S) = \{ a \in A \mid aS = 0 \} .$$

A right ideal is said to be a right annihilator if it has the form $r(S)$ for some $S \subset A$.

Proposition 17.4. If A satisfies ACC on right annihilators and is left coperfect, then A is semi-primary.

Proof. Consider the ascending chain of right annihilators

$r(J) \subset r(J^2) \subset \ldots$. By hypothesis we have $r(J^n) = r(J^{n+1})$ for some n . If $r(J^n) \neq A$, then the left A-module $A/r(J^n)$ has non-zero socle, which is of the form $I/r(J^n)$ for some left ideal $I \supset r(J^n)$. But the semi-simple module $I/r(J^n)$ is annihilated by J , so $JI \subset r(J^n)$, which gives $I \subset r(J^{n+1}) = r(J^n)$. Then $I = r(J^n)$, which is a contradiction, and hence $r(J^n) = A$, so $J^n = 0$.

References: Bass [7], Faith [27].

§ 18. Quasi-Frobenius rings

In this § we discuss three classes of rings: S-rings, PF-rings and QF-rings (in order of decreasing generality). An injective module E is a cogenerator if $\mathrm{Hom}(M,E) \neq 0$ for every module $M \neq 0$ (of course it suffices to consider cyclic modules M , so E is a cogenerator if and only if for every right ideal $I \neq A$ there exists $x \neq 0$ in E with $xI = 0$). In the following Proposition we let $E(A)$ denote the injective envelope of A_A .

Proposition 18.1. The following properties of A are equivalent:

(a) $E(A)$ is an injective cogenerator.

(b) Every simple right module is isomorphic to a minimal right ideal of A .

(c) $\mathrm{Hom}(C,A) \neq 0$ for every cyclic module $C \neq 0$.

(d) $l(I) \neq 0$ for every right ideal $I \neq A$.

(e) A has no proper dense right ideals.

Proof. (a) $\Rightarrow$ (b): If S is a simple module, there exists a non-zero $f: S \to E(A)$. Since A is essential in $E(A)$, the image of f must lie in A .

(b) $\Rightarrow$ (c): Clear, because C has a simple quotient module.

(c) $\Rightarrow$ (d): Obvious.

(d) $\Rightarrow$ (e): Immediate from Proposition 3.8.

(e) $\Rightarrow$ (a): A right ideal I is by definition dense if and only if $\mathrm{Hom}(A/I, E(A)) = 0$.

Definition. A is called a right S-ring if it has the properties of 18.1. (We have reversed the terminology of Morita [56], who calls these rings "left S-rings" and furthermore assumes minimum conditions on A).

Proposition 18.2. When A is a right self-injective ring, the following conditions are equivalent:

(a) A is a right S-ring.

(b) A_A is an injective cogenerator.

(c) $r(l(I)) = I$ for every right ideal I .

Proof. (a) $\Rightarrow$ (b): Clear, since A_A is an injective module.

(b) $\Rightarrow$ (c): Suppose $r(l(I)) \neq I$. Then there exists a non-zero homomorphism $f: r(l(I))/I \to A$. Since A is injective, the composed homomorphism $g: r(l(I)) \to r(l(I))/I \to A$ must be of the form $g(a) = ba$ for some $b \in A$. Since $g(I) = 0$, we have $b \in l(I)$. But then $g(a) = 0$ for every $a \in r(l(I))$, which is impossible.

(c) $\Rightarrow$ (a): Obvious by 18.1(d).

Definition. A is a right PF-ring if A_A is an injective cogenerator.

<u>Proposition 18.3.</u> The following properties of A are equivalent:

(a) A is a right PF-ring.

(b) A is right self-injective and an essential extension of its right socle, and A/J is semi-simple.

(c) Every faithful right module is a generator for Mod-A .

<u>Proof.</u> We will only prove the implication (b) ⇒ (a), since this is the only one we will need in the following. For the proof of (a) ⇒ (b) and (b) ⇔ (c) we refer to Azumaya [6] and Osofsky [60] (cf. also Kato [42]).

(b) ⇒ (a): We write A/J as a direct sum of indecomposable right ideals $\bar{I}_1, \ldots, \bar{I}_n$. We can lift each $\bar{I}_k$ to an indecomposable right ideal I_k of A so that $A = I_1 \oplus \ldots \oplus I_n \oplus I'$. Then $I' \subset J$ and hence $I' = 0$. Each I_k contains a unique minimal right ideal J_k . Every simple A-module S may also be considered as a simple A/J-module, and is therefore isomorphic to some $\bar{I}_k$. In particular, each of the minimal right ideals J_k must be isomorphic to some $\bar{I}_j$. If $J_i \cong J_k$, then $I_i \cong I_k$ since I_k is an injective envelope of J_k . It follows that the number of non-isomorphic right ideals J_k is equal to the number of non-isomorphic modules $\bar{I}_k$. Hence each $\bar{I}_k$, and then every simple module, is isomorphic to some J_k , and A is thus a right S-ring.

We next prove a very useful property of self-injective rings:

Proposition 18.4. The following properties of A are equivalent:

(a) Every homomorphism $f: I \to A$, where I is a finitely generated right ideal, has the form $f(a) = ca$ for some $c \in A$.

(b) A satisfies (i) $l(I_1 \cap I_2) = l(I_1) + l(I_2)$ for all finitely generated right ideals I_1, I_2.

(ii) $l(r(a)) = Aa$ for every $a \in A$.

Proof. (a) $\Rightarrow$ (b): If I_1 and I_2 are finitely generated right ideals, then obviously $l(I_1) + l(I_2) \subset l(I_1 \cap I_2)$. Suppose on the other hand that $a \in l(I_1 \cap I_2)$. We can define a homomorphism $\alpha: I_1 + I_2 \to A$ as

$$\alpha(b) = \begin{cases} b & b \in I_1 \\ (1+a)b & b \in I_2 \end{cases}$$

since these two expressions coincide on $I_1 \cap I_2$. By (a) there exists $c \in A$ such that $\alpha(b) = cb$. For $b \in I_1$ we thus have $cb = b$, i.e. $(c-1)b = 0$. As a consequence we may write $a = (c-1) + (1+a-c) \in l(I_1) + l(I_2)$, which proves (i).

For every $a \in A$ we have $Aa \subset l(r(a))$. If $b \in l(r(a))$, we can define a homomorphism $Aa \to Ab$ as $xa \mapsto xb$. It must be given by left multiplication with some $c \in A$, so in particular $b = ca$ and $b \in Aa$.

(b) $\Rightarrow$ (a): Consider $f: I \to A$ where I is finitely generated, say $I = a_1A + \ldots + a_nA$. We use induction on n, the case $n = 0$ being obvious. So we may assume there exist c and c' in A such that

$$f(a) = \begin{cases} ca & \text{for } a \in I' = a_1A + \ldots + a_{n-1}A \\ c'a & \text{for } a \in a_nA \end{cases}.$$

For $a \in I' \cap a_nA$ we must have $(c-c')a = 0$, so $c-c' \in$ $\in l(I' \cap a_nA) = l(I') + l(a_nA)$ by (i). We accordingly write

$c-c' = b-b'$ with $bI' = 0$ and $b'a_n = 0$. Then left multiplication by $c-b = c'-b'$ coincides with f on $I = I' + a_nA$.

Proposition 18.5. If A is right self-injective, then:

(i) $l(I_1 \cap I_2) = l(I_1) + l(I_2)$ for all right ideals I_1, I_2.

(ii) $l(r(I)) = I$ for every finitely generated left ideal I.

Proof. (i): The argument used in the preceding proof works equally well in the present situation.

(ii): Write $I = Aa_1 + \ldots + Aa_n$. Then $l(r(Aa_1 + \ldots + Aa_n)) =$
$= l(r(Aa_1) \cap \ldots \cap r(Aa_n)) = l(r(Aa_1)) + \ldots + l(r(Aa_n)) =$
$= Aa_1 + \ldots + Aa_n$ by (i) and 18.4(ii).

Definition. A is a QF-ring (or quasi-Frobenius ring) if it is both right and left artinian and right and left self-injective.

We will show, however, that much weaker conditions suffice to make A a QF-ring. Our first step is to show that the conditions may be made one-sided. For this we need the following reformulation of the definition:

Lemma 18.6. A right and left artinian ring is a QF-ring if and only if it satisfies

$$r(l(I)) = I \quad , \quad l(r(I')) = I'$$

for all right ideals I and left ideals I'.

Proof. A QF-ring satisfies the double annihilator conditions by 18.5. Conversely, these conditions imply that the operations r and l define an anti-isomorphism between the ordered sets of left resp. right ideals. They therefore reverse the lattice operations, so conditions (i) of 18.4 are satisfied, and hence the ring is right and left self-injective.

Proposition 18.7. If A is right or left artinian and is right or left self-injective, then A is a QF-ring.

Proof. There are two cases we must consider:

1) A is right artinian and right self-injective;

2) A is left artinian and right self-injective.

We can reduce case 1) to 2): If $I_1 \subset I_2 \subset \ldots$ is an ascending chain of finitely generated left ideals, then the descending chain $r(I_1) \supset r(I_2) \supset \ldots$ is stationary since A is right artinian. The ascending chain is then also stationary by 18.5, and A is thus left noetherian. But a ring which is both left noetherian and right artinian, is also left artinian.

We now consider case 2). Since A is left artinian, it is right coperfect and therefore right semi-artinian (Propositions 17.1 and 17.2). It then follows from 18.3 that A is a right PF-ring, and this implies $r(l(I)) = I$ for every right ideal I, by 18.2. We also have $l(r(I)) = I$ for every left ideal I by 18.5. By an argument similar to the previous reduction of 1) to 2), we finally have that A is right artinian, and by applying Lemma 18.6 we may conclude that A is QF.

We may weaken the conditions for A to be QF quite a bit more:

Proposition 18.8. If A is right or left noetherian and is right or left self-injective, then A is a QF-ring.

Proof. There are two cases to be considered:

1) A is left noetherian and right self-injective;

2) A is right noetherian and right self-injective.

Case 1): To show that A is left artinian, it suffices to show that A is semi-primary (cf. remark at the beginning of § 17). A/J is a left noetherian regular ring by 16.5, hence semi-simple. It remains to show that J is nilpotent. The ascending chain of two-sided ideals $r(J) \subset r(J^2) \subset \ldots$ is stationary since A is left noetherian. Let $r(J^n) = r(J^{n+1})$. From 18.5 follows that $J^n = J^{n+1}$, which implies $J^n = 0$ by the Nakayama lemma ([9], §6, Cor.2 de Prop.6).

Case 2): We prove a slightly stronger statement:

<u>Theorem 18.9.</u> If A satisfies ACC on right or on left annihilators and is right or left self-injective, then A is a QF-ring.

<u>Proof.</u> Case 1): Suppose A satisfies ACC on left annihilators and is right self-injective. Every finitely generated left ideal is a left annihilator by 18.5, so A is left noetherian. A is then QF by case 1) of 18.8.

Case 2): Suppose A satisfies ACC on right annihilators and is right self-injective. If $I_1 \supset I_2 \supset \ldots$ is a descending chain of finitely generated left ideals, then $r(I_1) \subset r(I_2) \subset \ldots$ is stationary by hypothesis, and so A is left coperfect and even semi-primary by 17.4. A is then a right PF-ring by 18.3, which implies that every right ideal is a right annihilator (18.2). A is therefore right noetherian. But since we have just proved that A is semi-primary, A must then be right artinian. We conclude from 18.7 that A is QF.

An interesting property of QF-rings is:

Proposition 18.10. Every module over a QF-ring is a submodule of a free module.

Proof. It suffices to show that every injective module is projective when A is QF. Since A is noetherian, every injective module is a direct sum of indecomposable injective modules, so it clearly suffices to show that $E(S)$ is projective, where S is any simple module. But S is isomorphic to a right ideal of A, and since A is self-injective, $E(S)$ is a direct summand of A.

Combining this result with § 11, Exercise 3, we find that over a QF-ring, a module is injective if and only if it is projective.

Example: The group ring of a finite group over a field is a QF-ring ([20], p. 402).

Exercises:

1. Show that an injective module is a cogenerator if and only if it contains a copy of each simple module.
2. Show that a right PF-ring with zero Jacobson radical cannot have essential right ideals $\neq A$ and must therefore be semi-simple. (This is a special case of the implication (a) $\Rightarrow$ (b) of 18.3, which we did not prove).
3. Show that if A is a right PF-ring and I is a maximal right ideal, then $l(I)$ is a minimal left ideal.

<u>References:</u> Azumaya [6], Björk [8], Eilenberg-Nakayama [23], Faith [27], Ikeda-Nakayama [37], Kato [42].

Chapter 5. Maximal and classical rings of quotients

§ 19. The maximal ring of quotients

We recall that a right ideal I of A is dense if $(I:a)$ has no left annihilators for any $a \in A$, and that the family $\underline{D}$ of dense right ideals is a topology, corresponding to the hereditary torsion theory cogenerated by $E(A)$. The ring $A_{\underline{D}}$ is the maximal right ring of quotients of A, and is denoted by Q_m.

Proposition 19.1. Q_m is its own maximal right ring of quotients.

Proof. This follows from Proposition 11.11 since $\underline{D}^e$ is the family of dense right ideals of Q_m.

Let H be the endomorphism ring of $E(A)$. We recall from § 8 that there is a commutative diagram

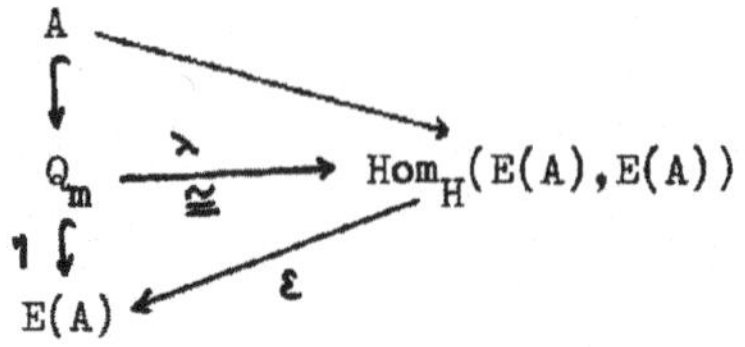

where λ is a ring isomorphism, η is the canonical inclusion (Proposition 8.2) and $\varepsilon(f) = f(1)$.

Proposition 19.2. The following assertions are equivalent:

(a) Q_m is a right self-injective ring.

(b) $\eta: Q_m \to E(A)$ is an isomorphism.

(c) The right ideal $(A:x)$ is dense for every $x \in E(A)$.

(d) There is a ring isomorphism $H \to Q_m$ such that the diagram

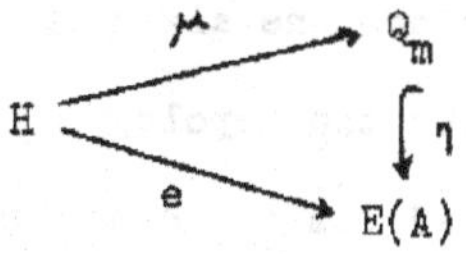

commutes, where $e(h) = h(1)$.

Proof. The equivalence of (b) and (c) is clear from Proposition 8.1, and the equivalence of (a) and (b) from Proposition 8.2.

(b)⇒(d): $Q_m \cong \mathrm{Hom}_{Q_m}(E(A),E(A)) = \mathrm{Hom}_A(E(A),E(A))$.

(d)⇒(b): e is obviously surjective, so if μ exists, then η is also surjective.

Lemma 19.3. If A is a right finite-dimensional ring, then so is also Q_m .

Proof. If a right ideal J of Q_m is a direct sum of non-zero right ideals J_α of Q_m , then each $J_\alpha \cap A$ is a non-zero right ideal of A . It follows that the family (J_α) must be finite.

Proposition 19.4. Suppose Q_m is right self-injective. Q_m is a semi-perfect ring if and only if A is right finite-dimensional.

Proof. $Q_m \cong \mathrm{Hom}_A(E(A),E(A))$ by 19.2, so Q_m is semi-perfect if and only if $E(A)$ is a finite-dimensional module, by 16.7. But $E(A)$ is finite-dimensional if and only if A is so.

Proposition 19.5. The following assertions are equivalent:

(a) $Q_m = M(A)$.

(b) Q_m is flat as a left A-module and $A \to Q_m$ is a ring epimorphism.

(c) Q_m is a right S-ring.

Proof. (a)$\Leftrightarrow$(b) is obvious since one always has $M(A) \subset Q_m$.

(a) $\Rightarrow$ (c): If $Q_m = M(A)$, then the topology $\underline{D}$ of dense right ideals of A is perfect. If J is a dense right ideal of Q_m , then $J \cap A \in \underline{D}$ by Example 3 of § 11. Therefore $Q_m = (J \cap A)Q_m \subset \subset J$, and $J = Q_m$.

(c) $\Rightarrow$ (a): If $I \in \underline{D}$ and $IQ_m \neq Q_m$, then $l(IQ_m) \neq 0$ by 18.1. Thus there exists $0 \neq q \in Q_m$ such that $qI = 0$. But this is impossible, because Q_m is $\underline{D}$-torsion-free.

Corollary 19.6. Suppose Q_m is right self-injective. Q_m is a right PF-ring if and only if $Q_m = M(A)$.

Proposition 19.7. Suppose Q_m is right self-injective. The following assertions are equivalent:

(a) Q_m is a quasi-Frobenius ring.

(b) Q_m is a Σ-injective right A-module.

(c) Every direct sum of $\underline{D}$-torsion-free injective A-modules is injective.

(d) The lattice $C_{\underline{D}}(A)$ of $\underline{D}$-pure right ideals is noetherian.

(e) A satisfies ACC on right annihilators of subsets of $E(A)$.

Proof. The equivalence of (b), (c), (d) and (e) was proved in Propositions 11.7 and 11.8. Since every QF-ring is noetherian, (a) $\Rightarrow$ (d) by Proposition 11.12.

(e) $\Rightarrow$ (a): By Theorem 18.9 it suffices to show that the ring Q_m satisfies ACC on right annihilator ideals. In view of condition (e), it will be enough to show that if S_1 and S_2 are subsets of Q_m, then $\{q \in Q_m \mid S_1q = 0\} \subset \{q \in Q_m \mid S_2q = 0\}$ if and only if $\{a \in A \mid S_1a = 0\} \subset \{a \in A \mid S_2a = 0\}$. The first inclusion trivially implies the second one. Suppose the second inclusion holds, and $S_1q = 0$ for some $q \in Q_m$. Choose $I \in \underline{D}$ such that $qI \subset A$. $S_1qI = 0$ implies $S_2qI = 0$, which gives $S_2q = 0$ since Q_m is torsion-free.

References: Lambek [47], Mewborn-Winton [54], Stenström [72].

§ 20. The maximal ring of quotients of a non-singular ring

For non-singular rings there are more precise results on the structure of Q_m than we were able to obtain in the general case. Recall that when A is right non-singular, the topology of dense right ideals coincides with the family of essential right ideals, which we denote by $\underline{E}$. Also the Goldie topology (§ 3, Example 1) reduces to $\underline{E}$ in this case.

Proposition 20.1. The following assertions are equivalent:

(a) A is right non-singular.

(b) Q_m is a regular ring.

(c) Q_m is a right self-injective regular ring.

(d) The Jacobson radical of $\mathrm{Hom}_A(E(A),E(A))$ is zero.

Proof. Suppose A is right non-singular. If $x \in E(A)$, then $(A:x)$ is essential. From Proposition 19.2 follows that Q_m is right self-injective and that $Q_m = \mathrm{Hom}_A(E(A),E(A))$. If we prove (a) $\Rightarrow$ (d), then also (a) $\Rightarrow$ (c) will follow, as a consequence of

Theorem 16.2.

(a) $\Rightarrow$ (d): Recall that the Jacobson radical of $\mathrm{Hom}(E(A),E(A))$ consists of those $\varphi: E(A) \to E(A)$ which have $\mathrm{Ker}\,\varphi$ essential in $E(A)$ (lemma 16.4). But if so is the case, then for each $x \in E(A)$ we have $xI \subset \mathrm{Ker}\,\varphi$ for some essential right ideal I, and $\varphi(x)I = 0$ implies $\varphi(x) = 0$ since $E(A)$ is non-singular; hence $\varphi = 0$.

(d) $\Rightarrow$ (a): Suppose $aI = 0$ where $a \in A$ and I is an essential right ideal. The homomorphism $A \to E(A)$ given by $b \mapsto ab$ extends to an endomorphism φ of $E(A)$. Then $\mathrm{Ker}\,\varphi \supset I$ is essential in $E(A)$, so φ is in the Jacobson radical of the endomorphism ring, and hence $\varphi = 0$. This implies $a = 0$.

(b) $\Rightarrow$ (a): Suppose again $aI = 0$ where $0 \neq a \in A$ and I is essential. We may write $a = aqa$ for some $q \in Q_m$. Since $qa \neq 0$ and I is essential in Q_m, there exists $b \in A$ such that $0 \neq qab \in I$. But then $ab = aqab \in aI = 0$ which contradicts $qab \neq 0$.

We see in particular from this Proposition that if Q_m is a semi-simple ring, then A is non-singular. Conversely we have:

<u>Theorem 20.2.</u> Let A be right non-singular. The following assertions are equivalent:

(a) Q_m is semi-simple.

(b) $Q_m = M(A)$.

(c) Q_m is flat as a left A-module and $A \to Q_m$ is a ring epimorphism.

(d) Every essential right ideal contains a finitely generated essential right ideal.

(e) A is right finite-dimensional.

(f) The lattice $\underline{C}_{\underline{E}}(A)$ of complemented right ideals is noetherian.

(g) Every direct sum of non-singular injective modules is injective.

(h) $E(A)$ is a Σ-injective module.

Proof. We know to begin with that Q_m is right self-injective and regular. Therefore the equivalence of (a), (f), (g) and (h) is immediate from Proposition 19.7. We have (a) $\Rightarrow$ (b) by 19.6, (b) $\Leftrightarrow$ (c) trivially, (c) $\Rightarrow$ (d) by Theorem 13.1.
We get (d) $\Leftrightarrow$ (e) from Propositions 19.4 and 19.3, and (d) $\Leftrightarrow$ (f) from Proposition 11.14.

Corollary 20.3. If Q_m is semi-simple, then:

(i) every right Q_m-module is injective as an A-module.

(ii) every left Q_m-module is flat as an A-module.

Proof. Follows from the formulas (1) and (2) of § 14.

Corollary 20.4. If Q_m is semi-simple and M is a non-singular A-module, then $M \otimes_A Q_m$ is an injective envelope of M.

Proof. We have $M_{\underline{E}} = E(M)$ by 8.1, and $M_{\underline{E}} = M \otimes_A Q_m$ by 13.1.

Proposition 20.5. Let A be right finite-dimensional non-singular. Q_m is then also a maximal left ring of quotients of A if and only if Q_m is flat as a right A-module.

Proof. If Q_m is a maximal left ring of quotients of A, then it is right flat over A by Theorem 20.2. On the other hand, if Q_m is right flat, then $A \to Q_m$ is a right flat epimorphism of rings. This implies that Q_m is a subring of $M_1(A)$ (i.e. the maximal right flat epimorphic extension ring of A) by Theorem 14.1. But $M_1(A)$ is a subring of the maximal left ring of quotients of A . Since Q_m is left self-injective, it must then coincide with the maximal left ring of quotients of A (cf. Proposition 19.1).

We will take a closer look at the lattices $\underline{C}_{\underline{E}}(M)$ of $\underline{E}$-pure submodules. We assume A to be right non-singular. As a special case of Proposition 11.5 we have:

Proposition 20.6. If M is a non-singular module, then $\underline{C}_{\underline{E}}(M)$ is a complemented modular lattice consisting of the complemented submodules of M .

We may describe $\underline{C}_{\underline{E}}(A)$ to a certain extent, because this lattice is isomorphic to the lattice of complemented right ideals of Q_m (Proposition 11.12). Since Q_m is right self-injective regular, this second lattice consists of the direct summands of Q_m . We thus have:

Proposition 20.7. If A is right non-singular, then $\underline{C}_{\underline{E}}(A)$ is isomorphic to the lattice of principal right ideals of Q_m .

Corollary 20.8. If A is right finite-dimensional non-singular, then $\underline{C}_{\underline{E}}(A)$ is a modular lattice of finite length.

Examples:

1. If A is a Boolean ring (i.e. a commutative regular ring with all elements idempotent), then Q_m is the "completion" of A in the sense of Boolean algebra theory.

2. It may happen that Q_m is flat as a left A-module even if Q_m is not semi-simple. In fact, one can just take A to be a regular ring which is not semi-simple. More generally:

Proposition 20.9. If A is right semi-hereditary, then A is right non-singular and Q_m is flat as a left A-module.

Proof. If $0 \neq a \in A$, then $A/r(a) \cong aA$ is a projective module, so $r(a)$ is a direct summand of A and can therefore not be essential in A. Hence A is non-singular.

To show that Q_m is flat over A, it suffices to show that the map $I \otimes Q_m \rightarrow A \otimes Q_m$ is a monomorphism for every finitely generated right ideal I. Since I is projective, $I \otimes Q_m$ is a projective Q_m-module and is therefore non-singular as a right A-module. Suppose $\sum a_i \otimes q_i \in I \otimes Q_m$ and $\sum a_i q_i = 0$. There is an essential right ideal J such that $q_i J \subset A$ for all i, and for each $a \in J$ we get $(\sum a_i \otimes q_i)a = \sum a_i(q_i a) \otimes 1 = 0$. $I \otimes Q_m$ non-singular implies $\sum a_i \otimes q_i = 0$.

Exercises:

1. Show that if A is regular and Q_m is projective as a right A-module, then $A = Q_m$. (Hint: an imbedding $E(A) \hookrightarrow F$, F free, induces $A \hookrightarrow F'$, F' finitely generated free; then use the fact that a finitely presented flat module is projective).

2. Let K be a skew-field and A the ring of upper triangular 2×2-matrices over K. Show that:

   (i) The matrices of the form

$$\begin{pmatrix} 0 & b \\ 0 & c \end{pmatrix}$$

   constitute a minimal essential right ideal of A.

   (ii) The ring $M_2(K)$ of all 2×2-matrices is the maximal right (and left) ring of quotients of A.

   (iii) $M_2(K)$ is projective as a right (and as a left) A-module.

References: Cateforis [15], [16], Faith [28], Lambek [47], Sandomierski [68], [69], Utumi [80], Walker and Walker [81].

§ 21. The maximal ring of quotients of a reduced ring

A ring is called reduced if there are no nilpotent elements $\neq 0$. In this § we want to find out when the maximal ring of quotients of a reduced ring also is reduced. For a commutative ring, this is always the case, because if $q \in Q_m$ and $q^n = 0$, then $0 \neq aq \in A$ for some $a \in A$ and also $(aq)^n = 0$.

Lemma 21.1. If A is reduced and $S \subset A$, then:

(i) $r(S)$ is a two-sided ideal.

(ii) $S \cap r(S) = 0$.

(iii) Every idempotent is central.

(iv) A is non-singular.

Proof. (i): It suffices to show that $r(S) = l(S)$. But $Sa = 0$ implies $aS = 0$ because $(aS)^2 = aSaS = 0$, and similarly $aS = 0$ implies $Sa = 0$.

(ii): $(S \cap r(S))^2 \subset Sr(S) = 0$, so $S \cap r(S) = 0$.

(iii): If $e^2 = e$ and $a \in A$, then $(ea(1-e))^2 = ea(1-a)ea(1-e) = 0$, so $ea = eae$. Similarly $ae = eae$, so $ea = ae$.

(iv): $r(a)$ cannot be essential in A since $aA \cap r(a) = aA \cap r(aA) = 0$ by (i) and (ii).

When A is commutative, the converse of (iv) holds, because a non-singular ring is a subring of a commutative regular ring, which obviously cannot have nilpotent elements $\neq 0$. In the non-commutative case, this is no longer true (consider e.g. matrix rings). Since A reduced implies Q_m regular, we should first determine under what conditions a regular ring is reduced.

Proposition 21.2. The following properties of a ring A are equivalent:

(a) A is a reduced regular ring.

(b) Every principal right ideal is generated by a central idempotent.

(c) A is regular and every right ideal is two-sided.

(d) A is strongly regular, i.e. for every $a \in A$ there exists $x \in A$ such that $a = a^2x$.

Proof. (a) $\Rightarrow$ (b): Clear from Lemma 21.1(iii).

(b) $\Rightarrow$ (c): Since every principal right ideal is two-sided, all right ideals are two-sided.

(c)$\Rightarrow$(d): For every $a \in A$ there exists x such that $a = axa$. Since axA also is a left ideal, $axa = bax$ for some b. Then $a^2 = axa \cdot a = baxa = ba$, so $a = a^2x$.

(d)$\Rightarrow$(a): A is obviously reduced since $a = a^2x$ implies $a = a^n x^{n-1}$ for each n. We also have $(a-axa)^2 = a^2 + axa^2xa - a^2xa - axa^2 = 0$, so $a = axa$ and A is regular.

If Q_m is a strongly regular ring, then A must be reduced. Conversely we have:

Proposition 21.3. The following properties of a reduced ring A are equivalent:

(a) Q_m is strongly regular.

(b) Every complemented right ideal of A is a two-sided ideal.

(c) $aA \cap bA = 0$ implies $ab = 0$, for all $a, b \in A$.

(d) $I \cap J = 0$ implies $IJ = 0$, for all right ideals I, J of A.

For the proof we need:

Lemma 21.4. If A is any ring with a topology $\underline{F}$ such that $A \subset A_{\underline{F}}$, then for each $I \in \underline{C}_{\underline{F}}(A)$ one has $I = IA_{\underline{F}} \cap A$.

Proof. Suppose $a = \Sigma a_i q_i \in A$ with $a_i \in I$ and $q_i \in A_{\underline{F}}$. Choose $J \in \underline{F}$ such that $q_i J \subset A$ for all q_i. Then $aJ = \sum a_i q_i J \subset I$, and $I \in \underline{C}_{\underline{F}}(A)$ implies $a \in I$.

Proof of 21.3: (a)$\Rightarrow$(b): Each complemented right ideal of A has the form $I = IQ_m \cap A$ by the Lemma. Since IQ_m is a two-sided ideal of Q_m , I is a two-sided ideal in A .

(b)⇒(d): If $I \cap J = 0$, let $K \supset J$ be a right ideal maximal with respect to $I \cap K = 0$. K is then two-sided, so $IJ \subset I \cap K = 0$

(d)⇒(c): Trivial.

(c)⇒(a): We know that Q_m is the endomorphism ring of E(A) (Proposition 19.2). Suppose $f: E(A) \to E(A)$ is nilpotent, say $f \neq 0$ but $f^2 = 0$. Since the Jacobson radical of the regular ring Q_m is zero, Ker f is not an essential submodule of E(A) . So there exists $0 \neq a \in A$ such that $aA \cap \text{Ker } f = 0$. But $f(a) \in \text{Ker } f$, so (c) implies that $f(a)a = 0$. Hence $a^2 \in \text{Ker } f$, and we must have $a^2 = 0$. Since A is reduced, this is a contradiction.

Proposition 21.5. Let A be reduced with Q_m strongly regular. Then:

(i) $\underline{C}_{\underline{F}}(A)$ is the family of right annihilator ideals.

(ii) $\underline{C}_{\underline{F}}(A)$ is a complete Boolean algebra with $I \mapsto r(I)$ as a complementation.

Proof. (i): Every ideal of the form $r(S)$, $S \subset A$, is complemented by Proposition 11.7. Suppose conversely I is complemented, say I is maximal with respect to $I \cap J = 0$. Then $I \subset r(J)$ by 21.3(d), and since also $r(J) \cap J = 0$ by 21.1, we must have $I = r(J)$.

(ii): A Boolean algebra is the same as a complemented distributive lattice [18]. That $r(I)$ is a complement of I follows from 21.1 and 21.3(d). Since $\underline{C}_{\underline{F}}(A)$ is isomorphic to the lattice of idempotents in Q_m, it suffices to show that this second lattice is distributive, i.e. $e\wedge(f\vee g) = (e\wedge f)\vee(e\wedge g)$. But $e\wedge f = ef$ and $e\vee f = e+f-ef$, and one has $e(f+g-fg) = ef+eg-efg$.

Example: Let A be strongly regular and right self-injective. A is then also left self-injective. For let B be the maximal left ring of quotients of A. B is then also strongly regular (e.g. by 21.3(b)). For each $b \in B$ there exists an idempotent $e \in A$ such that $0 \neq eb \in A$. Since every idempotent in B is central, we also have $0 \neq be \in A$, so A is an essential right submodule of B. Since A is right self-injective, B must equal A.

Exercises:

1. Show that every strongly regular ring is a subring of a product of skew-fields.
2. Let K be a skew-field and I an arbitrary set. Let A be the subring of $\prod_I K_i$ consisting of (x_i) such that almost all (i.e. all except a finite number) x_i are equal. Show:
   (i) A is strongly regular.
   (ii) $\bigoplus_I K_i$ is a minimal essential ideal of A.
   (iii) $\prod_I K_i$ is the maximal ring of quotients of A.

References: Renault [106], Utumi [113].

§ 22. The classical ring of quotients

When the classical ring of quotients Q_{cl} exists, it is a subring of Q_m, in fact a subring of $M(A)$. If Q_{cl} is right self-injective, we have $Q_{cl} = M(A) = Q_m$.

<u>Proposition 22.1.</u> The following properties of A are equivalent:

(a) Q_{cl} exists and is a right self-injective semi-perfect ring.

(b) A is right finite-dimensional and $(A:x)$ contains a non-zero-divisor for each $x \in E(A)$.

<u>Proof.</u> If Q_{cl} is right self-injective, then $Q_{cl} = E(A)$. Every $q \in Q_{cl}$ may be written $q = ab^{-1}$ where $a \in A$ and b is a non-zero-divisor in A. Hence $(A:q)$ contains a non-zero-divisor. If Q_{cl} furthermore is semi-perfect, then A is finite-dimensional by Proposition 19.4.

Suppose conversely that A is right finite-dimensional and $(A:x)$ contains a non-zero-divisor for every $x \in E(A)$. Every $(A:x)$ is then dense, because for any $a \in A$ we have $((A:x):a) = (A:xa)$, which has no non-zero left annihilators. Q_m is right self-injective and semi-perfect by Propositions 19.2 and 19.4. We will show that Q_m is a classical right ring of quotients of A. Since for each $q \in Q_m$ we have $qa \in A$ for some non-zero-divisor a of A, it suffices to show that every non-zero-divisor a of A is invertible in Q_m. Let $\varphi_a : Q_m \to Q_m$ be the map $q \mapsto aq$. Since the kernel of φ_a has zero intersection with the essential submodule A of Q_m, φ_a is a monomorphism. $\operatorname{Im} \varphi_a$ is thus an injective submodule of

Q_m , so $Q_m = \varphi_a(Q_m) \oplus K$. Iterating this, one gets $Q_m = \varphi_a^2(Q_m) \oplus \oplus \varphi_a(K) \oplus K$, using the fact that φ_a is a monomorphism. More generally, for each n one gets $Q_m = \varphi_a^n(Q_m) \oplus \varphi_a^{n-1}(K) \oplus \dots \oplus K$. Since A is finite-dimensional and $Q_m = E(A)$, we must therefore necessarily have $K = (0)$. φ_a is thus also an epimorphism, so a has a right inverse. To conclude the proof, we use:

<u>Lemma 22.2.</u> Suppose A has no infinite family of orthogonal idempotents. If $a \in A$ has a right inverse, then a is invertible.

<u>Proof.</u> Suppose $ab = 1$ but $ba \neq 1$. Define elements

$$e_i = b^i a^i - b^{i+1} a^{i+1} \quad \text{for } i = 1, 2, \dots .$$

Each e_i is different from zero, because $b^i a^i = b^{i+1} a^{i+1}$ would give $b^i a^i = ab^{i+1} a^{i+1} b = ab^i a^i b = b^{i-1} a^{i-1}$, using repeatedly the fact that $ab = 1$, and this would finally lead to $ba = 1$. One verifies immediately that the e_i are orthogonal idempotents, contrary to our assumption.

<u>Lemma 22.3.</u> If A satisfies ACC on right annihilators, then the singular right ideal is nilpotent.

<u>Proof.</u> We will show that the ascending chain $r(Z) \subset r(Z^2) \subset \dots$ would be strictly ascending if Z were not nilpotent. If $Z^n \neq 0$, choose an element $a \in Z$ with $Z^{n-1}a \neq 0$ and largest possible right annihilator. For each $b \in Z$ we have $r(b) \cap aA = 0$ since $r(b)$ is essential in A . So there exists $c \in A$ such that $ac \neq 0$ but $bac = 0$, which means that $r(ba)$ is strictly

larger than $r(a)$, and by the choice of a we must therefore have $Z^{n-1}ba = 0$. Since $b \in Z$ is arbitrary, we get $Z^n a = 0$, and hence $r(Z^{n-1})$ is strictly contained in $r(Z^n)$.

Proposition 22.4. Suppose A satisfies:

1) A is right finite-dimensional;

2) A satisfies ACC on right annihilators;

3) $(A:x)$ contains a non-zero-divisor for each $x \in E(A)$.

Then Q_{cl} exists and is a semi-primary right self-injective ring.

Proof. In view of Proposition 22.1 it remains to show that the Jacobsom radical J of Q_{cl} is nilpotent. From Lemma 16.4 we know that that J is the singular submodule of the right A-module Q_{cl} , and $Z(A) = J \cap A$. Lemma 22.3 says that $Z^n = 0$ for some n , and from this it follows that also $J^n = 0$. In fact, if $q_1, \ldots, q_n \in J$, write $q_i = a_i b_i^{-1}$ where a_i and b_i are in A and b_i are non-zero-divisors. Note that $a_i \in Z(A)$. We may write $q_1 \cdot \ldots \cdot q_n$ in the form $a_1' \ldots a_n' b^{-1}$, with $a_i' \in Z(A)$ and b a non-zero-divisor, by repeated use of the Ore condition. Hence $q_1 \ldots q_n = 0$.

By strengthening the condition (2) slightly we obtain a quasi-Frobenius classical ring of quotients.

Proposition 22.5. The following two assertions are equivalent:

(a) A satisfies

1) A is right finite-dimensional;

2) A satisfies ACC on right annihilators of subsets of $E(A)$;

3) $(A:x)$ contains a non-zero-divisor for each $x \in E(A)$.

(b) Q_{cl} exists and is a QF-ring.

Proof. (a)$\Rightarrow$(b): Q_{cl} exists and coincides with Q_m by 22.1. It is QF by 19.7.

(b)$\Rightarrow$(a): This follows also from Propositions 19.7 and 22.1.

We may now characterize those rings for which Q_{cl} is semi-simple ("Goldie's theorem").

Theorem 22.6. The following properties of A are equivalent:

(a) Q_{cl} exists and is a semi-simple ring.

(b) A is right finite-dimensional and right non-singular, and has no nilpotent two-sided ideals $\neq 0$.

(c) A is right finite-dimensional, satisfies ACC on right annihilators, and has no nilpotent two-sided ideals $\neq 0$.

(d) A is right finite-dimensional and every essential right ideal contains a non-zero-divisor.

Proof. (a)$\Rightarrow$(d): Since $Q_{cl} = Q_m$ is semi-simple, we know from Theorem 20.2 that for every essential right ideal I one has $Q_{cl} = IQ_{cl}$. Write $1 = \sum a_i q_i$ with $a_1, \ldots, a_n \in I$. Write $q_1 = b_1 c_1^{-1}$ where $b_1, c_1 \in A$ and c_1 is a non-zero-divisor. Then $c_1 = a_1 b_1 + \sum_2^n a_i q_i c_1$. Write $q_2 c_1 = b_2 c_2^{-1}$ with b_2,

$c_2 \in A$ and c_2 a non-zero-divisor. Then $c_1c_2 = a_1b_1c_2 + a_2b_2 + \sum_3^n a_iq_ic_1c_2$. Continuing in this manner, we get non-zero-divisors $c_1,\ldots,c_n$ such that $c_1\cdot\ldots\cdot c_n \in \sum a_iA \subset I$. Thus I contains a non-zero-divisor.

(d)$\Rightarrow$(a): Q_{cl} exists and is a right self-injective ring by 22.1. Since A is right non-singular and right finite-dimensional, it follows from Theorem 20.2 that $Q_{cl} = Q_m$ is semi-simple.

(a)$\Rightarrow$(c): From Proposition 22.5 follows that we only have to show that every nilpotent two-sided ideal I is zero. $l(I)$ is an essential right ideal, for if $a \neq 0$ is any element of A, let n be the smallest integer such that $aI^n = 0$; then there exists $b \in I^{n-1}$ such that $ab \neq 0$ and $ab \in l(I)$. But since (a)$\Leftrightarrow$(d), every essential right ideal contains a non-zero-divisor, and therefore I must be zero.

(c)$\Rightarrow$(b): $Z(A)$ is a nilpotent two-sided ideal by Lemma 22.3. By hypothesis it must be zero, so A is right non-singular.

(b)$\Rightarrow$(c): Q_m is semi-simple by Theorem 20.2 and this implies ACC on right annihilators by Proposition 19.7.

(c)$\Rightarrow$(d): We first prove:

<u>Lemma 22.7.</u> If A satisfies ACC on right annihilators and has no nilpotent two-sided ideals $\neq 0$, then every right or left nilideal is zero.

<u>Proof.</u> Since Aa is nil if and only if aA is nil, it suffices to consider a nilideal Aa. Assume $Aa \neq 0$. Among the non-zero elements of Aa, choose one $b \in Aa$ with maximal right annihi-

lator. For each $c \in A$, let $(cb)^k = 0$, $(cb)^{k-1} \neq 0$. Since $r(b) \subset r((cb)^{k-1})$, we must have equality by maximality. Hence $cb \in r(b)$ and $bAb = 0$, which gives $(AbA)^2 = 0$. Since every nilpotent ideal is zero, we get $b = 0$. This is a contradiction, and so $Aa = 0$.

We return to the proof of (c) $\Rightarrow$ (d): Let I be an essential right ideal. Since A has ACC on right annihilators and I is not a nilideal (by the Lemma), there exists $a_1 \neq 0$ in I such that $r(a_1) = r(a_1^2)$. If $I \cap r(a_1) \neq 0$, we continue and choose $a_2 \in I \cap r(a_1)$ such that $a_2 \neq 0$ and $r(a_2) = r(a_2^2)$. If then $I \cap r(a_1) \cap r(a_2) \neq 0$, we go on and get $a_3 \in I \cap r(a_1) \cap r(a_2)$, and so on. At each step we obtain a direct sum $a_1A \oplus \ldots \oplus a_kA$. This is proved by induction: suppose $a_1A \oplus \ldots \oplus a_{k-1}A$ is direct and $a_kb = a_1b_1+\ldots+a_{k-1}b_{k-1}$; since for each $i < k$ we have $a_ia_k = 0$, we get $b_i \in r(a_i^2) = r(a_i)$ and hence $\sum_1^{k-1} a_ib_i = 0$. But A is right finite-dimensional, so the process must stop at some stage, where we have $I \cap r(a_1) \cap \ldots \cap r(a_k) = 0$. Then $r(a_1) \cap \ldots \cap r(a_k) = 0$, so if $c = a_1+\ldots+a_k \in I$, then $r(c) = 0$. We need:

<u>Lemma 22.8.</u> If A is right finite-dimensional and $r(c) = 0$, then cA is an essential right ideal.

<u>Proof.</u> If I is any right ideal and $cA \cap I = 0$, then it is easy to see that we get a direct sum $I \oplus cI \oplus \ldots \oplus c^nI \oplus \ldots$.

We may now conclude the proof of the Theorem. Since A is right non-singular, it follows from the Lemma that $l(c) = 0$, and c is thus a non-zero-divisor belonging to I .

Exercise:

Let A be the ring of matrices

$$\begin{pmatrix} a & q \\ 0 & b \end{pmatrix}$$

with $a, b \in \mathbb{Z}$ and $q \in Q$. Show that the ring of upper triangular matrices over Q is a two-sided classical ring of quotients of A , while $M_2(Q)$ is the maximal right (and left) ring of quotients of A . (Note that A has nilpotent two-sided ideals $\neq 0$).

References: Gabriel [31], Goldie [32], Jans [39], Mewborn and Winton [54], Procesi and Small [64], Sandomierski [68].

REFERENCES:

1. J.S. Alin, Structure of torsion modules, Ph.D. Thesis, U. of Nebraska 1967.

2. J.S. Alin and S.E. Dickson, Goldie's torsion theory and its derived functor, Pacific J. Math. 24, 1968, 195-203.

3. G. Almkvist, Fractional categories, Arkiv f. Mat. 7, 1968, 449-476.

4. S.A. Amitsur, General theory of radicals II, American J. Math. 76, 1954, 100-125.

5. K. Asano, Uber die Quotientenbildung von Schiefringen, J. Math. Soc. Japan 1, 1949, 73-78.

6. G. Azumaya, Completely faithful modules and self-injective rings, Nagoya Math. J. 27, 1966, 697-708.

7. H. Bass, Finitistic dimension and a homological generalization of semi-primary rings, Trans. Amer. Math. Soc. 95, 1960, 466-488.

8. J.-E. Björk, Rings satisfying certain chain conditions, to appear.

9. N. Bourbaki, Algèbre, ch. 8, Hermann 1958.

10. - " - , Algèbre commutative, ch. 1 and 2, Hermann 1961.

11. B. Brainerd and J. Lambek, On the ring of quotients of a Boolean ring, Canad. Math. Bull. 2, 1959, 25-29.

12. I. Bucur and A. Deleanu, Introduction to the theory of categories and functors, J. Wiley 1968.

13. H. Cartan and S. Eilenberg, Homological algebra, Princeton 1956.

14. V.C. Cateforis, Flat regular quotient rings, Trans. Amer. Math. Soc. 138, 1969, 241-250.

15. V.C. Cateforis, On regular self-injective rings, Pacific J. Math. 30, 1969, 39-45.

16. - " - , Two-sided semisimple maximal quotient rings, Trans. Amer. Math. Soc. 149, 1970, 339-349.

17. S.U. Chase, Direct products of modules, Trans. Amer. Math. Soc. 97, 1960, 457-473.

18. P.M. Cohn, Universal algebra, Harper and Row 1965.

19. R.S. Cunningham, E.A. Rutter and D.R. Turnidge, Rings of quotients of endomorphism rings of projective modules, to appear.

20. C.W. Curtis and I. Reiner, Representation theory of finite groups and associative algebras, J. Wiley 1962.

21. S.E. Dickson, A torsion theory for abelian categories, Trans. Amer. Math. Soc. 121, 1966, 223-235.

22. M. Djabali, Anneau de fractions d'un J-anneau, Canad. J. Math. 17, 1965, 1041-1052.

23. S. Eilenberg and T. Nakayama, On the dimension of modules and algebras II, Nagoya Math. J. 9, 1955, 1-16.

24. V.P. Elizarov, On quotient rings of associative rings, Izv. Akad. Nauk SSSR 24, 1960, 153-170. (Amer. Math. Soc. Transl. 52, 1966).

25. - " - , Flat extensions of rings, Soviet Math. Dokl. 8, 1967, 905-907.

26. H. Eriksson, Fraktionkategorier och Grothendiecktopologier (mimeographed), Stockholm 1967.

27. C. Faith, Rings with ascending condition on annihilators, Nagoya Math. J. 27, 1966, 179-191.

28. - " - , Lectures on injective modules and quotient rings, Springer Lecture Notes 49, 1967.

30. C. Faith and E.A. Walker, Direct-sum representations of injective modules, J. Algebra 5, 1967, 203-221.

31. P. Gabriel, Des catégories abéliennes, Bull. Soc. Math. France 90, 1962, 323-448.

32. A.W. Goldie, Some aspects of ring theory, Bull. London Math. Soc. 1, 1969, 129-154.

33. O. Goldman, Rings and modules of quotients, J. Algebra 13, 1969, 10–47.

34. R.N. Gupta, Self-injective quotient rings and injective quotient modules, Osaka J. Math. 5, 1968, 69–87.

35. M. Hacque, Localisations exactes et localisations plates, Publ. Dép. Math. Lyon 6, 1969, 97–117.

36. G. Helzer, On divisibility and injectivity, Canad. J. Math. 18, 1966, 901–919.

37. M. Ikeda and T. Nakayama, On some characteristic properties of quasi-Frobenius and regular rings, Proc. Amer. Math. Soc. 5, 1954, 15–18.

38. J.P. Jans, Some aspects of torsion, Pacific J. Math. 15, 1965, 1249–1259.

39. – " – , On orders in quasi-Frobenius rings, J. Algebra 7, 1967, 35–43.

40. R.E. Johnson, Structure theory of faithful rings, I and II, Trans. Amer. Math. Soc. 84, 1957, 508–544.

41. A.I. Kašu, Closed classes of left Λ-modules and closed sets of left ideals of the ring Λ, Mat. Zametki 5, 1969, 381–390.

42. T. Kato, Self-injective rings, Tohoku Math. J. 19, 1967, 485–495.

43. – " – , Torsionless modules, Tohoku Math. J. 20, 1968, 234–243.

44. A.G. Kurosch, Radicals in rings and algebras, Mat. Sb. 33, 1953, 13–26.

45. J. Lambek, On the structure of semi-prime rings and their rings of quotients, Canad, J. Math. 13, 1961, 392–417.

46. – " – , On Utumi's ring of quotients, Canad. J. Math. 15, 1963, 363–370.

47. – " – , Lectures on rings and modules, Blaisdell 1966.

48. – " – , Torsion theories, additive semantics and rings of quotients, Springer Lecture Notes 177, 1971.

49. – " – , Bicommutators of nice injectives, to appear.

50. L. Levy, Torsion-free and divisible modules over non-integral domains, Canad. J. Math. 15, 1963, 132–151.

51. J.-M. Maranda, Injective structures, Trans. Amer. Math. Soc. 110, 1964, 98-135.

52. E. Matlis, Injective modules over noetherian rings, Pacific J. Math. 8, 1958, 511-528.

53. - " - , Modules with descending chain condition, Trans. Amer. Math. Soc. 97, 1960, 495-508.

54. A.C. Mewborn and C.N. Winton, Orders in self-injective semi-perfect rings, J. Algebra 13, 1969, 5-9.

55. A.P. Mishina and L.A. Skornjakov, Abelian groups and modules (in Russian), Moscow 1969.

56. K. Morita, On S-rings, Nagoya Math. J. 27, 1966, 687-695.

57. - " - , Localization in categories of modules I, Math. Z. 114, 1970, 121-144.

58. - " - , Localization in categories of modules II, J. Reine Angew. Math. 242, 1970, 163-169.

59. C. Năstăsescu and N. Popescu, On the localization ring of a ring, J. Algebra 15, 1970, 41-56.

60. B.L. Osofsky, A generalization of quasi-Frobenius rings, J. Algebra 4, 1966, 373-387.

61. - " - , Endomorphism rings of quasi-injective modules, Canad. J. Math. 20, 1968, 859-903.

62. N. Popescu and P. Gabriel, Caractérisation des catégories abéliennes avec générateurs et limites inductives exactes, C.R. Acad. Sci. Paris 258, 1964, 4188-4190.

63. N. Popescu and T. Spircu, Sur les épimorphismes plats d'anneaux, C.R. Acad. Sci. Paris 268, 1969, 376-379.

64. C. Procesi and L. Small, On a theorem of Goldie, J. Algebra 2, 1965, 80-84.

65. J.-E. Roos, Sur l'anneau maximal de fractions des AW^*-algèbres et des anneaux de Baer, C.R. Acad. Sci. Paris 266, 1968, 449-452.

66. - " - , On the structure of abelian categories with generators and exact direct limits. Applications, to appear.

67. Séminaire P. Samuel, Les épimorphismes d'anneaux, Paris 1968.

68. F.L. Sandomierski, Semisimple maximal quotient rings, Trans. Amer. Math. Soc. 128, 1967, 112-120.

69. - " - , Non-singular rings, Proc. Amer. Math. Soc. 19, 1968, 225-230.

70. L. Silver, Non-commutative localization and applications, J. Algebra 7, 1967, 44-76.

71. L.A. Skornjakov, Elizarov's quotient ring and the localization principle, Mat. Zametki 1, 1967, 263-268.

72. B. Stenström, On the completion of modules in an additive topology, J.Algebra 16, 1970, 523-540.

73. - " - , Flatness and localization over monoids, to appear in Math. Nachrichten.

74. H. Tachikawa, Double centralizers and dominant dimensions, Math. Z. 116, 1970, 79-88.

75. M.L. Teply, Torsion-free injective modules, Pacific J. Math. 28, 1969, 441-453.

76. - " - , Some aspects of Goldie's torsion theory, Pacific J. Math. 29, 1969, 447-459.

77. M.L. Teply and J.D. Fuelberth, The torsion submodule splits off, to appear.

78. D.R. Turnidge, Torsion theories and semihereditary rings, Proc. Amer. Math. Soc. 24, 1970, 137-143.

79. - " - , Torsion theories and rings of quotients of Morita equivalent rings, to appear.

80. Y. Utumi, On quotient rings, Osaka Math. J. 8, 1956, 1-18.

81. C.L. Walker and E.A. Walker, Quotient categories and rings of quotients, mimeographed, 1963.

82. L.E.T. Wu, H.Y. Mochizuki and J.P. Jans, A characterization of QF-3 rings, Nagoya Math. J. 27, 1966, 7-13.

83. O. Zariski and P. Samuel, Commutative algebra I, van Nostrand 1958.

Additional references:

84. T. Akiba, Remarks on generalized rings of quotients, III, J. of Math. Kyoto 9, 1969, 205-212.

85. E.P. Armendariz, On finite-dimensional torsion-free modules and rings, Proc. Amer. Math. Soc. 24, 1970, 566-571.

86. J.E. Björk, Rings satisfying a minimum condition on principal ideals, J. Reine Angew. Math. 236, 1969, 112-119.

87. A. Cailleau and G. Renault, Sur l'enveloppe injective des anneaux semi-premiers à l'idéal singulier nul, J. Algebra 15, 1970, 133-141.

88. V.P. Elizarov, Rings of quotients, Algebra i Logika Seminar 8, 1969, 381-424.

89. P. Gabriel and M. Zisman, Calculus of fractions and homotopy theory, Springer 1967.

90. A.W. Goldie, The structure of prime rings under ascending chain conditions, Proc. London Math. Soc. 8, 1958, 589-608.

91. - " - , Semi-prime rings with maximum condition, Proc. London Math. Soc. 10, 1960, 201-220.

92. R.N. Gupta, Self-injective quotient rings and injective quotient modules, Osaka J. Math. 5, 1968, 69-87.

93. N. Jacobson, Structure of rings, Amer. Math. Soc. Coll. Publ. 37, revised ed. 1964.

94. R.E. Johnson, Prime rings, Duke Math. J. 18, 1951, 799-809.

95. - " - , Extended centralizer of a ring over a module, Proc. Amer. Math. Soc. 2, 1951, 891-895.

96. - " - , Quotient rings of rings with zero singular ideal, Pacific Math. J. 11, 1961, 1385-1392.

97. - " - , Rings with zero right and left singular ideals, Trans. Amer. Math. Soc. 118, 1965, 150-157.

98. R.E. Johnson and E.T. Wong, Quasi-injective modules and irreducible rings, J. London Math. Soc. 36, 1961, 260-268.

99. J.T. Knight, On epimorphisms of non-commutative rings, Proc. Cambr. Phil. Soc. 68, 1970, 589-601.

100. D. Lazard, Autour de la platitude, Bull. Soc. Math. France 97, 1969, 81-128.

101. A.C. Mewborn, Some conditions on commutative semiprime rings, J. Algebra 13, 1969, 422-431.

102. B. Mitchell, Theory of categories, Academic Press 1965.

103. K. Morita, Localization in categories of modules III, Math. Z. 119, 1971, 313-320.

104. N. Popescu and T. Spircu, Quelques observations sur les épimorphismes plats (à gauche) d'anneaux, J. Algebra 16, 1970, 40-59.

105. N. Popescu and D. Spulber, Sur les quasi-ordres (à gauche) dans un anneau, J. Algebra 17, 1971, 474-481.

106. G. Renault, Anneaux réduits non commutatifs, J. Math. Pures Appl. 46, 1967, 203-214.

107. - " - , Anneau associé à un module injectif, Bull. Sci. Math. 92, 1968, 53-58.

108. J.C. Robson, Artinian quotient rings, Proc. London Math. Soc. 17, 1967 , 600-616.

109. J.E. Roos, Locally distributive spectral categories and strongly regular rings, Reports Midwest Category Seminar, Springer Lecture Notes 47, 1967, 156-181.

110. L. Small, Orders in artinian rings, J. Algebra 4, 1966, 13-41 and 505-507.

111. H. Tachikawa, Localization and artinian quotient rings, Math. Z. 119, 1971, 239-253.

112. M. Takeuchi, A simple proof of Gabriel and Popescu's theorem, J. Algebra 18, 1971, 112-113.

113. Y. Utumi, On rings of which any one-sided quotient rings are two-sided, Proc. Amer. Math. Soc. 14, 1963, 141-147.

114. E.T. Wong and R.E. Johnson, Self-injective rings, Canad. Math. Bull. 2, 1951, 167-173.